I0464638

Authors

Janaki Alavalapati is professor and head, Department of Forestry, College of Natural Resources, Virginia Polytechnic Institute and State University, Blacksburg, VA 24061. **Ralph J. Alig** is a research forester, U.S. Department of Agriculture, Forest Service, Pacific Northwest Research Station, 3200 SW Jefferson Way, Corvallis, OR 97331. **Pankaj Lal** is a Ph.D. Candidate, School of Forest Resources and Conservation, University of Florida, Gainesville, FL 32611. **D. Evan Mercer** is a research economist, U.S. Department of Agriculture, Forest Service, Southern Research Station, P.O. Box 12254, Research Triangle Park, NC 27709. **Anita T. Morzillo** is an assistant professor, senior research, Department of Forest Ecosystems and Society, College of Forestry, Oregon State University, Corvallis, OR 97331. **Edward A. Stone** is a graduate research assistant, Department of Agricultural and Resource Economics, Oregon State University, Corvallis, OR 97331. **Eric M. White** is a research associate, Department of Forest Engineering, Resources and Management, College of Forestry, Oregon State University, Corvallis, OR 97331.

Cover photograph by Ralph J. Alig.

Effects of Climate Change on Natural Resources and Communities: A Compendium of Briefing Papers

Ralph J. Alig, Technical Coordinator

U.S. Department of Agriculture, Forest Service
Pacific Northwest Research Station
Portland, Oregon
General Technical Report PNW-GTR-837
March 2011

Abstract

Alig, Ralph J., tech. coord. 2011. Effects of climate change on natural resources and communities: a compendium of briefing papers. Gen. Tech. Rep. PNW-GTR-837. Portland, OR: U.S. Department of Agriculture, Forest Service, Pacific Northwest Research Station. 169 p.

This report is a compilation of four briefing papers based on literature reviews and syntheses, prepared for USDA Forest Service policy analysts and decisionmakers about specific questions pertaining to climate change. The main topics addressed here are effects of climate change on wildlife habitat, other ecosystem services, and land values; socioeconomic impacts of climate change on rural communities; and competitiveness of carbon offset projects on nonindustrial private forests in the United States. The U.S. private forest offset projects tend to be less costly than European projects but more expensive to implement than those in tropical forests in developing countries. Important policy considerations involving any mitigation actions include effects on other ecosystem services, such as wildlife habitat, and determining baselines and additionality. Stacking of ecosystem services payments or credits with carbon offset payments may be crucial in improving the participation of private forest owners. Potential social impacts of climate change are discussed in terms of health effects on rural communities and climate change sensitivity of indigenous communities. Potential economic impacts on rural communities are discussed for agriculture, forestry, recreation and tourism, fisheries, water resources, and energy. Salient findings from the literature are summarized in the synthesis of the literature, along with identified research needs.

Keywords: Climate change, wildlife habitat, land values, ecosystem services, vulnerability, rural communities.

Summary

Periodic syntheses of the ever-expanding knowledge about climate change, impacts on natural resources and associated ecosystems, land use and values, and human communities are critical to effective policy formulation. We also need to better understand how adaptation to and policies for addressing climate change may affect landscapes, ecosystems, ecosystem services, and local economies. Policy deliberations will be aided by better information on the economic viability of forest-related offsets relative to other greenhouse gas (GHG) offset options. Forest ecosystems can transfer carbon from the air as part of the GHG complex and sequester it into plant tissue through the process of photosynthesis during the growth of trees and in other ecosystem components such as the understory and soil. Such forest sinks have a significant potential to help in mitigating climate change, and this report is a compilation of briefing papers prepared for USDA Forest Service policy analysts and decisionmakers about specific forest carbon sequestration topics. The briefing papers are part of a larger set prepared by agency scientists and cooperators, including a related one regarding forest bioenergy by White[1] and another set pertaining to economic modeling of the effects of climate change on the U.S. forest sector and mitigation options.[2] Given the large topic of climate change and forests, this report only touches on selected aspects, and readers are referred to the growing literature for information on specific topics such as incorporating climate change considerations into specific natural resource management.[3]

In the first chapter, Morzillo and Alig review the literature pertaining to effects of climate change on wildlife and wildlife habitat at a broad scale. Because climate change is such a far-reaching topic, research opportunities focusing on climate change impacts on wildlife and wildlife habitat are likely to be most informative for policymakers if formulated using an integrated and multidisciplinary systems analysis approach and adaptive management. An adaptive process will facilitate

---

[1] White, E. 2010. Woody biomass for bioenergy and biofuels in the United States—a briefing paper. Gen. Tech. Rep. PNW-GTR-825. Portland, OR: U.S. Department of Agriculture, Forest Service, Pacific Northwest Research Station. 45 p.

[2] Alig, R. 2010. Economic modeling of effects of climate change on the forest sector and mitigation options: a compendium of briefing papers. Gen. Tech. Rep. PNW-GTR-833. Portland, OR: U.S. Department of Agriculture, Forest Service, Pacific Northwest Research Station. 169 p.

[3] Joyce, L.A.; Haynes, R.W.; White, R.; Barbour, R.J., tech. coords. 2006. Bringing climate change into natural resources management: proceedings. Gen. Tech. Rep. PNW-GTR-706. Portland, OR: U.S. Department of Agriculture, Forest Service, Pacific Northwest Research Station. 150 p.

results from research involving land-use changes, forest management, wildlife management, mitigation alternatives, and society's willingness to undertake actions to formulate more policy alternatives, along with confronting other components of the global climate issue.

The second chapter by Alig et al. assesses research about effects of climate change on land productivity and values, including influences on provision of ecosystem services. Climate change policies may in turn affect land and resource markets, thereby modifying land values, land use, and forest cover distribution. A broad examination of such changes necessitates assessing multiple markets, including those for carbon sequestration as an ecosystem service, especially if policies arise to promote carbon sequestration as a mitigation activity.

In the third chapter, Lal et al. look at effects of climate change on rural communities. Their review suggests that rural communities tend to be more dependent on climate-sensitive livelihood activities and have fewer resources and social support systems compared to urban populations. They suggest that rural communities could face large potential impacts from future climate change events.

In the fourth chapter, Mercer et al. review competitiveness of forest-based carbon offset projects, relative to other offset options in agriculture, clean transportation, carbon capture and sequestration, nuclear and other advanced technologies, increasing energy efficiency, renewable energy, and other options. They examine estimates of per unit costs of removing atmospheric carbon through domestic offset projects, and find that estimates differ widely.

Contents

Chapter 1: Climate Change Impacts on Wildlife and Wildlife Habitat

Anita T. Morzillo and Ralph J. Alig

Introduction

Climate change is affecting environmental conditions for people and wildlife. As a result of changing environmental conditions, climate change affects wildlife species both directly and indirectly. Current policies influence the area and distribution of potential wildlife habitat. In response to climate change, policies adopted to influence the management of ecosystems are expected also to affect the wildlife habitat and populations that depend on these ecosystems (IPCC 2007, MEA 2005). Many researchers have explored potential impacts of climate change on wildlife populations and habitats, as well as shifts in vegetation communities resulting from projected climate change. The focus of this paper is to provide a broad-scale review and synthesis of existing knowledge about potential direct and indirect impacts of climate change on terrestrial wildlife and wildlife habitat within a systems analysis context. To do this, we rely on existing research that has suggested climate-influenced impacts on particular species and their habitats. Although the readers of this document may have primary interest in U.S. policy, we use examples from around the world to illustrate phenomena that have potential to be relevant to wildlife within the United States.

Many studies have addressed observed and projected changes in land cover and land use as a result of climate change (e.g., Backlund et al. 2008, Harsch et al. 2009, Iverson et al. 2008, Shafer et al. 2001, White et al. 2010). For our purposes, we define land cover as the observed biophysical cover of the Earth's surface (e.g., oak-hickory forest, grassland). In contrast, land use is the means by which land is used by humans (e.g., protected areas, timberland, agriculture). Research focused on the impacts of climate change on wildlife, such as those discussed in this paper, expand on earlier studies focused on land cover and land use change to illustrate and project how organisms that depend on particular attributes of land cover and land use may respond to changes. Wildlife species depend on vegetation characteristics for cover (e.g., shelter from predators), reproduction (e.g., trees for denning and nesting), and food (e.g., pollen, seeds, and fruits). We describe direct ecological and behavioral responses of wildlife to climate change, such as geographical range shifts by species. Many studies have focused on location-specific case studies (e.g., Moritz et al. 2008) or shifts by individual or groups of organisms (e.g., Chen et al. 2009), but others provide a comprehensive analysis across many species and locations (e.g., Hughes 2000, Lawler et al. 2009, Parmesan 2006, Thomas and Lennon 1999). Although it is not our objective to quantitatively evaluate net change across all species, or provide an exhaustive review, we summarize existing literature by focusing on both spatial and temporal aspects of range shifts, and the biological effects of such shifts (Root et al. 2003).

The next section illustrates impacts of climate change on wildlife habitat. Here we focus on impacts that ecosystem changes in vegetation characteristics and associated abiotic variation may have on the species that depend on them. Topographical landscape variation (e.g., continental drift and other geologically induced change) may be slow in many cases from a human perspective. However, other effects of abiotic factors and combined abiotic and biotic influences may be more apparent (e.g., Harsch et al. 2009). Habitat variation as a result of climate change is expected to affect individual species differently, based on life history characteristics and ability to adapt to changing habitat conditions. We describe impacts of climate change mitigation on wildlife habitat from mainly an economic perspective, such as decisions that landowners may make in response to climate change. For example, potential return from emerging carbon and bioenergy markets may affect landowner decisions about land management, which may affect availability of habitat for particular species.

To integrate the first three parts of the paper, we describe and follow linkages between models for evaluating climate change and land use impacts on wildlife, including global circulation models, ecological process models, economic models, ecosystem services models (e.g., wildlife habitat), climate (or bioclimatic) envelope models, and attendant feedbacks, and critique the linkages to identify needs for better understanding of potential impacts. Finally, we suggest focal areas for future research to address climate change impacts, specifically the need to evaluate concurrent influences of climate change and land use on wildlife and wildlife habitat.

Ecological and Behavioral Responses of Wildlife to Climate Change

Adjusting to climate-influenced variation in resources they depend on is a challenge for wildlife. This variation may include changes in spatial distribution in food resources, timing of food resource availability, and differences in cover, ground moisture, and water requirements.

Range Shifts

Many abiotic (e.g., physical barriers, climate) and biotic (e.g., competition, population dynamics, life history, genetics) factors individually, and in combination, affect the geographic distribution of species (Gaston 2003). Researchers are observing variation over space and time in species distributions that appear to be range shifts in response to climate change. Geographical range shifts may be latitudinal, which are defined as extinction of a species at the historically observed southern (in the Northern Hemisphere; northern in the Southern Hemisphere) boundary or net colonization by the species at the northern (or southern in Southern Hemisphere) boundary (Parmesan et al. 1999). Parmesan and Yohe (2003) completed a meta-analysis of 1,700 species to evaluate consistencies between recent biological trends in species range distributions and climate change predictions. For wildlife, this analysis included both timing of life history events (phenology) and distributional changes in taxa including insects and other invertebrates, fish, reptiles and amphibians, birds, and mammals. Globally, on average across all studies evaluated, northern and upper elevation boundaries are estimated to have shifted approximately 6 km per decade toward the poles, which follows trends in climate models (Parmesan and Yohe 2003). The same authors suggested that changes in distributions of approximately 280 species followed a systematic trend in global cooling and warming periods since 1930.

Besides changes along latitudinal gradients, geographic range shifts also may occur along elevation gradients. The same meta-analysis that suggested many range distributions to be shifting toward the poles also noted elevation shifts upward of approximately 6 m per decade (Parmesan and Yohe 2003). In Costa Rica, variation in sea surface temperatures is influencing forest mist frequency, which then affects vertebrate abundance along elevation gradients that cannot be attributed to other land uses (Pounds et al. 1999). From fixed plots, observations of breeding populations of species typically found in areas lower than approximately 1500 m have increased in frequency above 1540 m. For example, the formerly lowland-breeding keel-billed toucan (*Ramphastos sulfuratus*) now breeds among the cloud-forest-breeding quetzals (*Pharomachrus moccino*) (Pounds et al. 1999). Moths (Chen et al. 2009) and butterflies also have exhibited elevational changes in distribution. Ranges of populations of Edith's checkerspot (*Euphydryas editha*) and sachem skipper butterfly (*Atalopedes campestris*) in North America (Crozier 2003, 2004; Parmesan 1996), as well as Apollo butterfly (*Parnassius apollo*) and other species (Descimon et al. 2005, Parmesan 1996, Parmesan and Yohe 2003, Wilson et al. 2005), have shifted upward along elevational gradients as compared to historical ranges.

Similar upward-elevation trends have been observed with high-altitude-associated mammals. Several pika (*Ochotona princeps*) populations across the Western United States are suspected to have gone extinct since first observed in the 1930s (Beever et al. 2003). In a followup evaluation, Parmesan and Galbraith (2004) suggested that extinct populations had occurred at significantly lower elevations

D. Canning

Climate-induced microhabitat changes that affect small mammal and bird populations may also affect predators that depend on such populations for food.

than surviving populations. A re-creation of Grinnell's survey of a 3000-m elevation gradient in Yosemite National Park also suggested elevational shifts among mammal communities. For example, approximately half of high-elevation species, including the alpine chipmunk (*Tamias alpinus*) and Belding's ground squirrel (*Spermophilus beldingi*), experienced a narrowing of elevational limits (decreased range sizes), whereas approximately half of low-elevation species (e.g., pocket mouse [*Chaetodippus californicus*] expanded their upper elevation maximum (Moritz et al. 2008).

Several long-term studies across broader taxonomic groups also report range shift dynamics. For example, in Britain, the northern margins of many breeding bird species distributions have moved north by 19 km during a 20-year period (Thomas and Lennon 1999). Similar trends have been reported for butterflies. Parmesan et al. (1999) observed that ranges of 22 out of 35 (63 percent) species of nonmigrating European butterflies have shifted north by 35 to 240 km since 1900. Comparable shifts have been reported for North America and for different habitat types (NABCI 2010) (fig. 1-1). According to the National Audubon Society, Christmas Bird Count results suggest that more than half (170 out of 305) of commonly observed species are shifting their wintering ranges northward by an average of 35 miles (NABCI 2009, 2010; The Audubon Society 2010) (fig. 1-2). These species include the purple finch (*Carpodacus purpureus*), wild

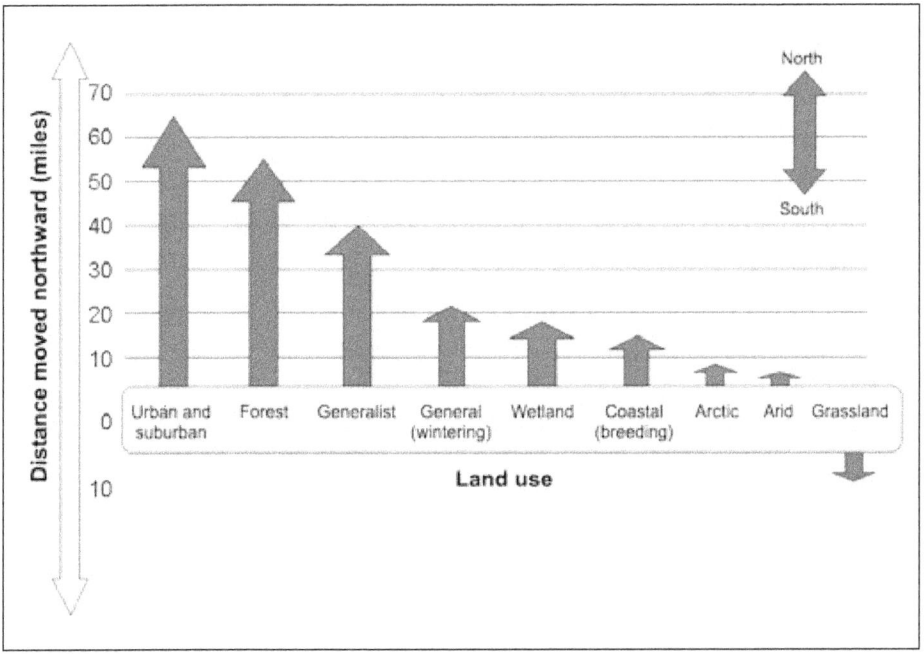

Figure 1-1—Forty-year (1960s–2006) latitudinal distance moved north by North American bird species across different land use types as observed from Christmas Bird Count data. The greatest shifts were observed among species associated with urban and suburban habitats. Adapted from NABCI 2010.

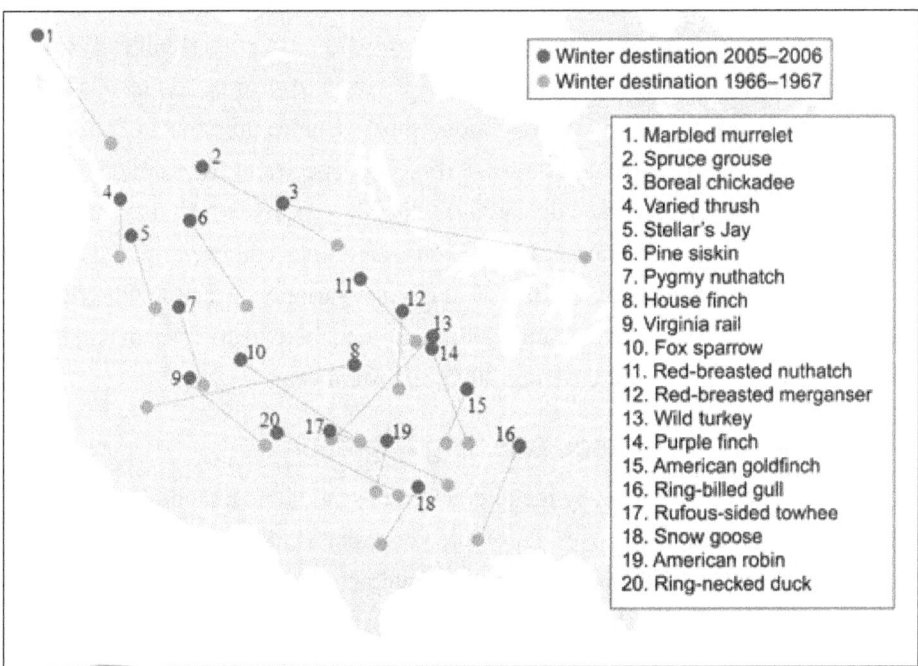

Figure 1-2—As the temperature warms across North America, the winter habitat of many birds is shifting northward. Dotted lines indicate approximate changes in wintering locations of 20 bird species. Adapted from The Associated Press and Audubon Society 2010.

turkey (*Meleagris gallopavo*), marbled murrelet (*Brachyramphus marmoratus*), boreal chickadee (*Poecile hudsonicus*), and house finch (*Carpodacus mexicanus*) (The Audubon Society 2010). The U.S. Committee of the North American Bird Conservation Initiative has reported that American robins (*Turdus migratorius*), tree swallows (*Tachycineta bicolor*), red-winged blackbirds (*Agelaius phoeniceus*), and eastern bluebirds (*Sialia sialis*) are breeding at least 1 week earlier than 30 years ago (NABCI 2009). In contrast, populations of other species, including the burrowing owl (*Athene cunicularia*) and eastern and western meadowlark (*Sturnella magna* and *S. neglecta*, respectively), may continue to decline as climate affects habitat quality, and lack of adequate habitat limits the ability of species to move northward (The Audubon Society 2010).

Temporal range shifts also exist in response to climate change. Temporal changes may be gradual, and indicate wildlife response to individual climate variables, such as variation in spring temperature. For example, the North Atlantic Oscillation may affect arrival dates for migratory bird species in New England (Wilson 2007). Although temporal cyclical climate phenomena may be attributed to behavior, other research has suggested noncyclical temporal changes. In Massachusetts, between 1970 and 2002, a 2 °C increase in average temperatures was accompanied by earlier timing of phenological events for 22 species including wood duck (*Aix sponsa*), ruby-throated hummingbird (*Archilochus colubris*), ovenbird (*Seiurus aurocapillus*), and chipping sparrow (*Spizella passerine*) (Ledneva et al. 2004). Similar results were observed for birds during a 61-year period in Wisconsin (Bradley et al. 1999), and particularly among short-distance migrants in Poland (Tryjanowski et al. 2002). In central New York, four species of frogs are calling 10 to 13 days earlier than observed during the early 20th century, which may signal a potential shift toward earlier reproductive behavior (Gibbs and Breisch 2001). It also is suspected that a combination of temporal climate variation affecting ultraviolet radiation and environmental contamination may contribute to declines in amphibian populations (Beebee and Griffiths 2005, Blaustein et al. 2001).

Environmental Processes Affecting Migration

Wildlife also may be affected by the interaction of spatial and temporal characteristics of environmental processes. Discussion of climate change often focuses on impacts on ecological systems. However, the interaction between ecological and abiotic systems characteristics may hinder a species' ability to adapt to ecological changes. Deviation of concurrent timing of abiotic and biotic environmental events may result in species vulnerability. Annual photoperiod (i.e., sunlight per 24 hours) cycles are relatively constant, and not related to climatic shifts. In other words, the same date each year has approximately the same amount of sunlight regardless of climatic events. Thus, innate behaviors triggered by length of daylight period (Gwinner 1996) may not be affected by climate change. Migratory birds often depend on changes in photoperiod, temperature, and wind to determine timing of spring and fall migrations (Gwinner 1996, Price and Glick 2002). Similarly, migratory butterflies use sensors in their antennae to determine timing for migration based on photoperiod (Kyriacou 2010). In short, particular ecological and behavioral events have become synchronized with photoperiod characteristics. Consequentially, species that migrate based on photoperiod characteristics have become adapted to using resources in winter and summer ranges that are available when a particular species arrives in the respective range. As ecological communities adapt to climate change, species behavior based on photoperiod likely will become out of synch with ecosystem processes in different parts of an organism's range. It is unknown whether wildlife will be able to adapt to asynchrony between photoperiod and ecosystem processes.

Although the preceding paragraphs have focused mainly on flying migratory species, it is important to note that climate change also impacts nonflying migrants. Many factors influence the ability of wildlife to migrate over land, including variation in forage quantity and quality, snow depth, ability to use particular ("traditional") areas and migration routes, and water availability (Harris et al. 2009). Long-term changes in any one of these properties may affect migration success. For example, forage moisture and access to water are important factors in resource quality for American pronghorn (*Antilocapra americana*) and influences migration behavior of this species (Harris et al. 2009). Plant biomass, including protein content, affects caribou selection of calving grounds and potential nutritional intake of both females and calves (Griffith et al. 2002). In Alaska and Canada, changing plant phenology has resulted in earlier visits to calving grounds by caribou (*Rangifer tarandus*) since the mid-1990s, and misplacement of forage availability combined with predation risk could result in a decrease in caribou herd size (Griffith et al. 2002). In support of such speculation, results of a long-term study (1988–2003) of caribou on the Québec-Labrador peninsula suggest that both climate and habitat variables have affected annual spatial and temporal movement patterns of and habitat use by caribou (Sharma et al. 2009). Regional corridor networks have been suggested as a means to maintain migration routes within changing landscapes (Berger 2004), but such efforts require a large amount of societal cooperation at the local level.

From a societal perspective, presence of mass-migrating species plays an important role in local and regional economies. For example, wildlife-related tourism is an important industry in Africa, and absence of mass-migrating wildlife (e.g., wildebeests [*Connochaetes taurinus*] and the predators that rely on them) could result in great impacts to regional tourism economies. Similarly, caribou hunting in the Arctic is not only an economic activity, but also a source of sustenance for native societies (Berkes and Jolly 2001, Sharma et al. 2009). Loss of the aforementioned economic sectors may force dependent locals to switch to more traditional yet environmentally intensive practices such as livestock and agriculture (Harris et al. 2009). In contrast, concern about migration routes could affect regional economic activity such as development. For example, various types of development (e.g., energy development) may affect migration routes and potentially the health of migratory wildlife populations such as mule deer (*Odocoileus hemionus*) and pronghorn (Sawyer et al. 2002, 2007), and uncertainty exists about how climate change may affect such migration routes and related relationships with particular land uses (Harris et al. 2009). Ultimately, tradeoffs exist between human resource use and mass-migrating species, particularly if wildlife show fidelity to particular migratory routes and locations that may be affected by climate change.

Functional Roles Within Ecosystems

There are important implications of climate change on wildlife beyond survival of the species themselves, such that displacement of land-migrating species as well as other wildlife may affect the greater ecosystem. For example, large ungulates are critical for driving ecosystem functions such as enriching grassland production and replenishing soil nutrients (Harris et al. 2009). In Africa, browsing by ungulates plays an important role in maintaining cooperative relationships between nectar-feeding ants and *Acacia* spp. trees by limiting the presence of antagonistic ant species (Palmer et al. 2008). However, climate-induced ecosystem changes are not limited to effects on larger ungulates, but may extend to other prey species. Increasingly warm early springs in Norway may be influencing humidity and snow characteristics that affect cyclical population irruptions of lemmings (*Lemmus lemmus*) (Kausrud et al. 2008). It is speculated that loss of lemming population irruptions already may be having profound impacts on nutrient cycling, functioning of plant groups, and predator populations (e.g., arctic fox [*Alopex lagopus*] and snowy owl [*Bubo scandiacus*]) (Kausrud et al. 2008). In northern New York, projected changes in spring runoff may affect distribution of beaver (*Castor canadensis*) populations and associated impounded wetlands, the presence of which is depended upon by mink frogs (*Rana septentrionalis*) for mating and reproduction (Popescu and Gibbs 2009). Ultimately, the climate-related impacts on one species may have cascading effects on other species within an ecosystem.

No consensus exists on which wildlife species are most vulnerable to effects of climate change.

Researchers have not reached consensus on which wildlife species are most vulnerable to the effects of climate change. Some suggest that long-distance migratory species may be particularly vulnerable because of their dependence on the interaction between abiotic cues and ecological resources, as described in this section (NABCI 2009, 2010; Price and Glick 2002; Price and Root 2005; Robinson et al. 2009). However, nonmigrating leaf-feeding insects and other herbivores also can exhibit phenological synchrony with host plants (van Asch and Visser 2007). Others suggest that species with limited climatic ranges (e.g., alpine species such as pikas) or restricted reproductive strategies or physiologies may be most at risk (Issac 2009, Tewksbury et al. 2008). For example, many small mammals such as mice (e.g., *Peromyscus* spp.) are capable of breeding multiple times per year, whereas many larger mammals (e.g., black bears, *Ursus americanus*) reproduce on an annual cycle only. Regardless of vulnerability, researchers speculate that climate change may affect morphological and genetic species traits (Issac 2009, Root et al. 2003), which has potential to affect species-habitat relationships. There is a great need to better understand the spatial and temporal adjustments that wildlife will have to make to climate-induced habitat changes (described in the next section), which ultimately is a concern for all wildlife.

Impacts of Climate Change on Wildlife Habitat

The previous section highlighted ecological and behavioral responses of wildlife to climate change. However, climate change impacts to wildlife also may include disturbances that drive changes in wildlife habitat. Examples of climate-influenced changes that may indirectly affect wildlife are variation in vegetation community composition or other ecosystem components, as well as pest outbreaks and other disturbances such as wildfire. Some species may be relatively more sensitive to overall habitat change than other species. For example, amphibians have received a large amount of recent attention in that the presence of additive habitat-related elements (e.g., chytrid fungus [*Batrachochytrium dendrobatidis*]) has had potentially major impacts on their populations. Other factors that may affect habitat, such as the interaction between land use and climate change, are only now becoming a focus of research.

Relationships Between Vegetation and Wildlife

Vegetation type, often identified as a main characteristic of land cover (e.g., evergreen forest), is an important predictor of species occurrence. Temperature (Körner 2007) and interaction between position and temperature (Körner and Paulsen 2004) often are considered primary drivers of plant species distribution, although other factors also may be influential, such as precipitation, disturbance, and community interactions (Harsch et al. 2009). Globally, many researchers have suggested that plant communities are shifting toward the poles, upward in elevation, and blooming earlier in the season, any of which likely will affect the wildlife that depend on them.

Research has identified relationships between treeline range shifts and climate variables. In a meta-analysis, Harsch et al. (2009) evaluated a global data set of 166 sites across more than 100 studies for which treeline dynamics since 1900 could be classified; most sites were in North America or Europe. The authors observed recruitment of treelines at increasingly higher altitudes or latitudinal advancement of treelines at 87 out of 166 sites, which corresponded with sites that also exhibited warming during the winter months. Treelines with diffuse form (gradual merge between tree communities or to treeless areas) were strongly associated with changes in annual and winter temperatures. In contrast, shift of abrupt and krummholz (stunted) treelines were more associated with winter warming and were potentially affected by other constraints on tree survival (e.g., wind, ice) (Harsch 2009). At a regional level, 78 percent of 542 leafing, flowering, and fruiting records across 21 European countries were observed to have advanced in latitude since 1971 (Menzel et al. 2006). Along with trees, herbs, and shrubs, even symbiotic organismal groups such as lichens are susceptible to range shifts that would be projected by climate change (Parmesan and Yohe 2003).

Vegetation range shifts that may affect the distribution of habitat also may be influenced by interactions between climatic variables and disturbance. In New Mexico, Allen and Breshears (1998) reported a ≥2-km shift in ponderosa pine (*Pinus ponderosa* Dougl. ex Laws.) and piñon-juniper woodland over a 5-year period, a result of several interrelated climate-influenced events. In the same study, a combination of drought, fire suppression, and a bark beetle outbreak may have been factors in a concurrent expansion of the piñon-juniper community. Piñon-juniper eventually may have competed with the ponderosa pines for limited water, which resulted in enhanced susceptibility of the ponderosa pines to bark beetle attack and related mortality (Allen and Breshears 1998). Ultimately, climate change impacts on plant communities could affect temporal aspects of resource use and consumption by dependent wildlife species, and survivorship of dependent wildlife populations.

Integration of climate model simulations with bioclimatic variables suggest shifts in vegetation taxa to continue with anticipated changes in climate. Shafer et al. (2001) suggested that projected range changes for particular tree and other species may be multidirectional, and could result in range fragmentation, both of which may affect associated wildlife. In an analysis by Iverson et al. (2008), a simulation modeling approach was used to project suitable habitat for 134 tree species across the Eastern United States. Results suggest that approximately 66 species would gain suitable habitat (e.g., southern oaks and pines), whereas 54 species would lose at least 10 percent of their current suitable habitat (e.g., spruce-fir). Range shifts were predicted to move up to several hundred miles in a generally northeastern direction. From the same study, the authors concluded that proactive means to lower greenhouse gases may limit range disruptions (Iverson et al. 2008).

Disturbance

It is expected that climate change will affect the frequency, extent, duration, and severity of disturbances, but a challenge exists in projecting specific outcomes (Dale et al. 2000). Individual disturbances may affect landscape structure or composition, which in turn affect wildlife habitat (Inkley et al. 2004). Even though disturbance is part of the natural landscape, climate-influenced variation of disturbance occurrence may affect the ability of wildlife to respond and adapt to individual disturbances.

Insect infestations within forest ecosystems is one type of disturbance that is believed to be affected by climate change. It has been speculated that warmer winters and summers and reduced summer precipitation have facilitated mountain pine beetle (*Dendroctonus ponderosae*) outbreaks in western Canada (Kurz et al. 2008) and the expansion of this pathogen's range into previously uninfected forests and different forest communities (Logan and Powell 2001). In Alaska, a combination of higher than average summer temperatures, increased winter survival of the spruce beetle pathogen, shortened pathogen maturation rate (2 years to 1 year), regional drought, and an adequate number of spruce trees all were likely contributors to a massive spruce beetle outbreak during the 1990s (Berg et al. 2006). The same authors suggested that continued warming trends may result in beetle populations being great enough to infect trees as soon as trees are mature enough to be susceptible to the insects. Insect infestations and resulting tree mortality have been estimated to cause as much as twice the diminution to forests as other disturbances in the boreal forests of Canada (Volney and Flemming 2000). As such, climate-induced projected increases in infestation occurrence and range (Williams and Liebhold 2002) are expected to have potential to cause changes to not only wildlife habitat, but also other forest processes such as carbon sequestration (Volney and Flemming 2000).

Anita Morzillo

Changes in both climate and habitat variables, such as those that affect forage moisture and water availability, have potential to affect migration behavior of pronghorn.

The number and severity of forest fires are expected to be affected by climate change. Flannigan et al. (2000) used two general circulation climate models to project climatic effects on forest fire severity. Forest fire severity was projected to increase between 10 and 50 percent for most of the United States, but details of the projections differed between climate models. Integrating projections from both climate models, the same authors suggested a majority of increases in fire severity may be expected in Alaska and the Southeastern United States, whereas other climate variables may result in lower expected fire severity in other locations (e.g., northern Great Plains). It is important to note that many locally affected variables (e.g., fuel load and type) were not included in these projections, and inclusion of them may lead to potential forecasting variation at a local level (Flannigan et al. 2000). Ultimately, fire occurrence likely will affect community composition and result in variation in habitat available for particular wildlife species.

Climate-affected disturbance also may influence the ability of exotic species to invade and potentially outcompete native vegetation. In southeastern Texas, Harcombe et al. (1998) suggested that frequency and intensity of natural disasters, such as hurricanes, floods, and droughts, may have more profound effects on changes to the vegetation community than climate-induced changes on individual species growth rates. From the same study, results of long-term observation indicate that presence of Chinese tallow (*Triadica sebifera* (L.) Small.) or *Sapium sebiferum* (L.) Roxb.), which establishes itself in gaps within the forest, has increased in population size by a factor of 30 between 1981 and 1995. This range expansion may be attributed to the fast-growing Chinese tallow outcompeting other species (Harcombe et al. 1998), a process that is expected to continue to spread as winter freeze extent and severity shift northward. If the frequency of natural large-scale disaster disturbances increases the frequency of forest disturbance, then fast-growing exotics such as Chinese tallow may interrupt establishment of native forest communities and reduce regeneration of plant species depended upon by wildlife.

Microclimate

Climate-related impacts present special conservation challenges for particularly abiotic-sensitive species such as amphibians (Carey and Alexander 2003), which may serve as indicators of environmental stress (Blaustein and Wake 1995). As noted in the preceding section, such species are known to respond to stresses initiated by regular oscillations such as El Niño (Pounds and Crump 1994). For example, an increase in amount of ultraviolet radiation as a result of climate variation has been observed to affect survivorship of amphibian eggs and embryos (Kiesecker et al. 2001). Amphibians also are susceptible to changes in water availability, which has been associated with warming-influenced dry periods (Pounds et al. 1999). Consequently, progressive climate-influenced factors may be additive when other stressors act on amphibian populations simultaneously. For example, a combination of ultraviolet radiation and toxins (e.g., pesticides) may enhance the potential of each to contribute to amphibian mortality (Blaustein et al. 2003) (fig. 1-3). Some researchers have suggested that climate change may play a role in creating better growing conditions for chytrid fungus, which has been attributed to the decline of mass extinctions of amphibians (Pounds et al. 2006). However, others argue that the causal link between chytrid fungus and global warming is not well supported at this time (Rohr et al. 2008). In fact, the long-term observation of declines in amphibian and lizard species in Costa Rica suggest that the chytrid fungus, which is specific to amphibians, may be an additive factor of mortality in addition to the greater potential threat of climate change (Wake 2007). In the same study, the authors suggested that trees are retaining leaves longer, and decomposition rates may be faster, both of which decrease potential refuge for wildlife on the forest floor and decrease an important environmental component for amphibians and reptiles. However, Wake (2007) suggested the need to test such inferences further.

Interaction Between Climate Change and Land Use

Climate change impacts also may be enhanced by concurrent human-caused drivers of habitat change. Many researchers have evaluated direct effects of disturbance on particular species resulting from land use. For example, Laliberté et al. (2010) observed an inverse relationship between land use intensification and both diversity among traits representing plant function (e.g., leaf size, nutrient uptake strategy) and response to disturbance (e.g., resprouting ability, age of reproduction), which suggests that land use intensification may decrease the resiliency (ability to successfully respond to stressors) of the overall affected ecosystem. If plant resiliency decreases, then it is possible that dependent wildlife will be increasingly vulnerable during disturbances.

Ultimately, however, few studies evaluate concurrent interactions between climate change and land use and resulting effects on wildlife habitat. Warren et al. (2001) observed that a majority of 46 British butterfly species that may have responded positively, via range expansion, to climate-induced changes suffered negative responses to habitat loss from development. Ultimately, habitat generalists adapted more successfully to both disturbances than habitat specialists (Warren et al. 2001). In another example, bird community composition change on Cape Cod since 1930 was suggested to be attributed more to global warming than to

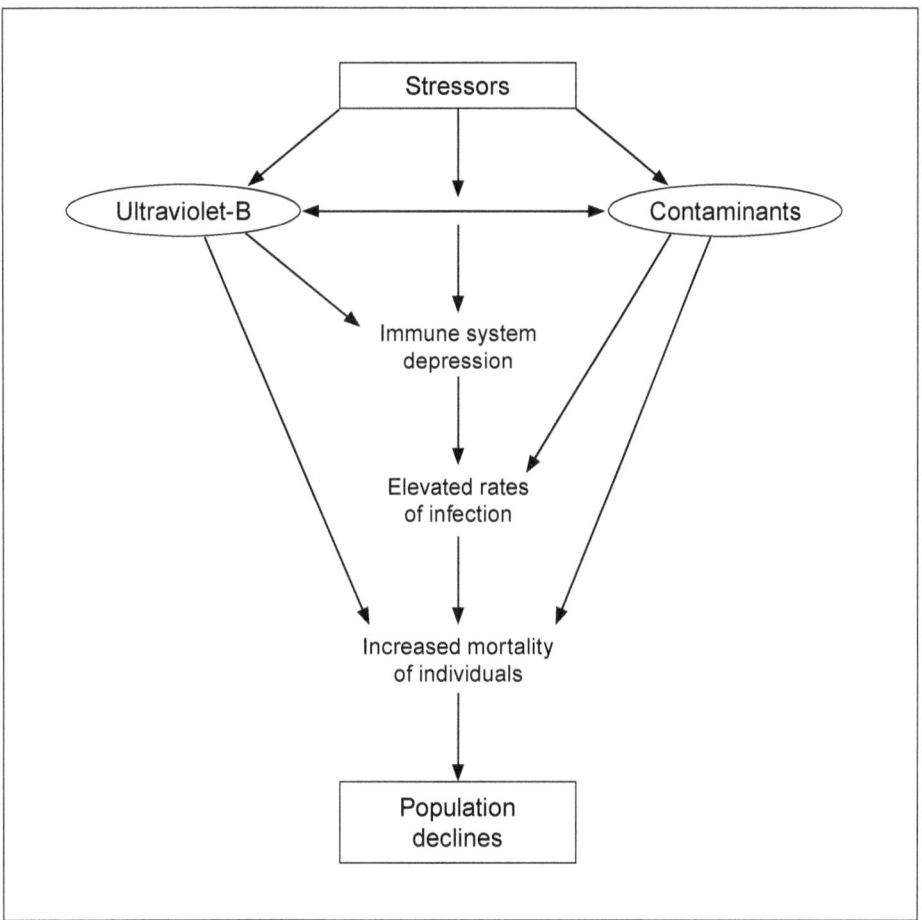

Figure 1-3—The simultaneous interactions between biotic and abiotic factors or stressors affecting amphibians. Either ultraviolet-B (UV-B) radiation or contaminants alone may affect mortality. For example, UV-B may depress the immune system, leaving species more susceptible to infection. Contaminants and UV-B radiation interacting together could heighten lethal effects. Ultimately, individual stressors or combinations of stressors may lead to declines in amphibian populations. Adapted from Blaustein et al. 2003.

urbanization (Valiela and Bowen 2003). From the same study, observations included increase in southern-associated species across all local habitat types, and decrease in northern-associated species since 1970. Although changes in northern-associated species since 1970 may be attributed to land cover change (increase in edge habitat from urbanization), collective changes were more associated with temperature increases than land use. McAlpine et al. (2009) argued the need to evaluate interactions between climate and land cover, and the importance of vegetation in maintenance of regional climatic conditions. Focusing on the Australian continent, the authors suggest that historical clearing for agriculture (approximately 15 percent of the continent) and other land use changes that resulted in removal of native vegetation not only reduced ecosystem resilience but also may have contributed to climate changes across the continent. Such drastic change may have drastic implications for wildlife at the continental level, but this assertion has not been tested. Although beyond the scope of this paper, it is speculated that both climate change and land use change may have profound effects on sediment load in streams within the Alps, which in turn may affect brown trout (*Salmo trutta*) habitat (Scheurer et al. 2009). In the end, potential interactions between climate change and land use effects on wildlife may suggest a need to consider both factors in planning processes and mitigation policy (McAlpine et al. 2009).

Impacts of Climate Change Mitigation on Wildlife Habitat

As landscape conditions respond to climate change, resource managers will be forced to respond to the new landscape conditions as well as any related climate change policy. Mitigation for climate change, defined as policies or actions that seek to decrease the impact of climate change, likely will influence decisions about land use. In turn, wildlife habitat likely will be affected by decisions about land use such as afforestation, carbon and bioenergy markets, and incentive programs for landowners. Interactions between mitigation and adaptation activities also will play a role, although very little currently is known about such interactions and will not be discussed further here.

Afforestation

Policies supporting afforestation may affect amount of habitat available for wildlife, as well as habitat quality. Increased area and quality of wildlife habitat, particularly for forest species, is a possible co-benefit of conversion of land use from row-crop agriculture to forest land. Matthews et al. (2002) used econometric models of land use to simulate the response of landowners to planting trees on land currently used for agriculture in South Carolina, Maine, and southern Wisconsin. These models then were used in combination with breeding bird survey data to evaluate responses of forest and farmland bird populations. For a scenario of 10-percent conversion of agriculture for carbon sequestration, projected land conversion between agriculture and forest differed in amount from state to state. Following these projected changes in land use, results from Matthews et al. (2002) also suggested overall losses of farmland birds, and only small increases in forest bird species. However, from the same study, overall total bird populations were projected to decrease between 1 and 3 percent across all three states. The authors suspected that these differences were effects of overlaps between econometric predictions of locations converted and spatial patterns of bird species richness. Ultimately, tradeoffs between biodiversity and economic benefits will need to be evaluated by policymakers and landowners.

Other research has evaluated the potential benefits for afforestation, such as conservation of wildlife habitat, which may be derived from carbon sequestration programs in the upper Midwestern United States (Feng et al. 2007, Plantinga and Wu 2003). Decreased soil erosion, nitrogen pollution, and atrazine pollution were among additional benefits that could be derived from the afforestation program, in addition to improvement of wildlife habitat (Plantinga and Wu 2003). From the same study, the authors suggested that overall benefits derived from the program would be the same order of magnitude as the costs associated with the carbon sequestration program. Feng et al. (2007) expanded earlier work by suggesting that particular benefits and transfers of them may vary geographically.

Although afforestation may offer a means to provide additional habitat for wildlife, particularly forest species, unintended consequences that affect other ecosystem components are possible. In Ireland, policy was initiated to double the percentage of forest land by 2035 (Allen and Chapman 2001). From the same study, thus far, afforestation efforts and resulting update and storage (by trees and soil) of water have led to a reduction in precipitation runoff by as much as 20 percent. However, new

Gary Wilson, Natural Resources and Conservation Service

Expanded afforestation as part of climate change mitigation could provide additional habitat for wildlife.

forestry operations have the potential to enhance risk of flooding within the area. In addition, a slower rate of groundwater recharge occurred, along with concerns about water quality (Allen and Chapman 2001). As another example, Thompson et al. (2009) suggested that afforestation efforts may be a factor in absorption of more solar radiation by forests and result in local warming temperatures. This local warming may neutralize or counter efforts to offset climate change by afforestation (Thompson et al. 2009). As such, the authors suggested a geographically targeted approach that seeks to increase forest land in locations naturally occupied by forests.

Carbon and Bioenergy Markets

The emergence of natural resource markets, such as for carbon and bioenergy, may result in changes for wildlife habitat. Ultimately, however, negative impacts of market-associated activities to some species likely will result in positive impacts to others. From an economic perspective, participating landowners may seek to maximize net financial returns from land management activities. For example, because older trees typically can store more carbon, some landowner decisions in response to carbon markets may include longer rotation times between timber harvests (White et al. 2010). An increase in rotation times between forest harvests could result in a decrease in habitat for early- to mid-succession wildlife species. At the same time, longer rotation times also may be beneficial for species that prefer relatively older growth forest habitat. Likewise, a decrease in number of timber harvests over time may result in decreased forest fragmentation.

Emergence of bioenergy markets also may result in both positive and negative impacts to wildlife habitat. Fargione et al. (2009) suggested that impacts from bio-fuels production run on a continuum between low and high wildlife habitat benefits (table 1-1). An example from the continuum suggests that cropland may hold less potential benefit for wildlife than diverse native habitats. Plant community research supports the notion that greater plant diversity is directly related to community productivity (Spehn et al. 2005, Tilman et al. 2001) and carbon storage (Tilman et al. 2001), and that limited diversity can affect other nonplant species and trophic levels within the ecosystem (Knops et al. 1999). Other research speculates that biomass feedstock may have a greater negative impact on wildlife if planted in marginal cropland than if planted within its respective geographic range with associated wildlife species that depend on it as a habitat resource (Fargione et al. 2009).

Although positive and negative effects of biofuels feedstock production on wildlife habitat have been identified conceptually (Bies 2006), research results are starting to emerge that focus on site-specific examples. Walter et al. (2009)

Table 1-1—Factors influencing wildlife habitat value and impacts on wildlife as related to bioenergy crops

Factors	Better for wildlife	Worse for wildlife
Habitat type	Diverse native habitats	Cropland
Plant diversity	Diverse native grasslands/forests	Exotic monocultures
Invasiveness of planted material	Native, noninvasive	Nonnative, invasive
Harvest and disturbance timing	Late fall, early spring	Breeding or nesting season
Harvest frequencies	Single harvest not more than once per year	Multiple harvests per year
Stubble height postharvest	Tall stubble or regrowth	Little or no stubble
Habitat refugia	Unharvested area within field	No unharvested area nearby
Landscape context	Complex of habitat patches	Isolated habitat patches
Type of land use replaced with biomass	Marginal cropland	Native habitat
Fertilizer use	Minimal	High
Pesticide use	Minimal	High
Soil erosion and sedimentation	Perennial plants and low erosion	Annual plants and high erosion

Adapted from Fargione et al. (2009).

evaluated impacts of conversion of cropland to native grassland, forest land, and aquatic vegetation on white-tailed deer (*Odocoileus virginianus*) within the DeSoto National Wildlife Refuge in eastern Nebraska. During the study, cropland (corn, soybeans, sorghum, alfalfa, and wheat-clover mix) was reduced by 44 percent between 1991 and 2004. Changes in plant communities did not affect overall landscape use by deer, but deer did shift home ranges to access remaining cropland (Walter et al. 2009). The authors suggested that such home range shifting may influence disease transmission rates as a result of new interactions between deer social groups (Walter et al. 2009).

Incentive Programs for Landowners

There are many means by which private landowners may derive benefits from their land while managing land for wildlife habitat. For example, hunting leases may provide forest-land owners extra income while also promoting biodiversity (Johnson 1995). Researchers also have assessed economic benefits of landowners adopting "green" practices, such as conservation tillage (Kurkalova et al. 2006). More broadly and formally, incentive programs may serve as a vehicle to entice private landowners to manage land in ways that are beneficial for habitat of particular wildlife species. For example, incentives created by the U.S. Department of Agriculture Conservation Reserve Program (CRP) have influenced availability of wildlife habitat. Establishment of more than 985,000 acres (398 615 ha) of grasslands coincided with a large increase in Henslow's sparrow (*Ammodramus henslowii*) populations in Illinois (Herkert 2007). Similarly, Niemuth et al. (2007) projected that loss of CRP lands would result in large declines in grassland bird populations among the prairie pothole region of North and South Dakota. Other incentive programs, such as the CRP Longleaf Pine Initiative, target a particular vegetation species. This program seeks to restore up to 250,000 acres (101 171 ha) of longleaf pine forests in nine Southern States, the benefit of which includes enhancement of native wildlife habitat (USDA 2006) and may also contribute to carbon sequestration.

As another example, the Wetlands Reserve Program is a voluntary program managed by the USDA Natural Resources Conservation Service (NRCS) that offers landowners the opportunity to protect, restore, and enhance wetlands on their property (USDA NRCS 2010). The NRCS goal is to achieve the greatest wetland functions and values, including wildlife habitat, and the agency provides technical and financial support to help landowners with wetland restoration efforts. Thus, this program offers landowners an opportunity to establish long-term conservation and wildlife practices and protection. There are three enrollment options: (1) permanent easement, (2) 30-year easement, and (3) restoration cost-share agreements. The 2008 Farm Bill changed the process for determining the easement value, increased the number of acres that could be enrolled, and set guidelines for payments and landownership tenure. By 2008, approximately 2 million acres (809 371 ha) nation-wide were enrolled in the Wetlands Reserve Program. In the next chapter, Alig et al. discuss how such programs and climate change programs could influence value of ecosystems services and land values, thereby influencing landowner behavior regarding land conservation and climate change mitigation.

It is suspected that private lands will play an important role in future biodiver-sity conservation. Two-thirds of all watersheds that are associated with privately owned forest land contain at least one at-risk species that is listed under the U.S. Endangered Species Act or ranked globally (Robles et al. 2008, Stein et al. 2010). The same studies show that many of the watersheds with the greatest number of species and highest density of each species are located in states within the southeast and within Appalachia, as well as along the Pacific coast. These locations also align with the highest proportion of estimated private forest conversion during the next 20 years (Robles et al. 2008). The authors suggested conservation easements and other incentive programs as potential vehicles to maintain biodiversity through landowner forest management programs.

Besides forest management programs, other competing land uses such as agriculture may be targeted for maintenance of wildlife habitat and promotion of biodiversity. From an economic perspective, Feng et al. (2006) evaluated alterna-tive approaches to conservation on the same parcel of land: retirement of land from production versus a change in farming practices on working land. The authors concluded that changes to working land were more cost-effective to the landowner if the objective is low levels of environmental benefits, such as a single benefit of erosion reduction, carbon sequestration, or wildlife habitat. However, land retirement was suggested to be more cost-effective if multiple benefits were sought, such as a combination of the aforementioned benefits (Feng et al. 2006). Plantinga (1996) also evaluated tradeoffs between keeping marginal farmland in production or conversion to forest to increase other environmental benefits, and suggested that many of the derived benefits from retiring agricultural land would result in enhancement of public goods (e.g., wildlife habitat, scenery, carbon sequestration, reduced soil erosion). However, Nelson et al. (2008) suggested that tradeoffs will exist between meeting objectives of policies designed to promote carbon sequestration and habitat conservation in heterogeneous landscapes.

Anita Morzillo

Impacts of climate change on timing of production of seasonal forage such as buds, insects, flowers, and berries may affect the ability of species such as the ptarmigan (*Logopus* spp.) to meet food resource needs, particularly during young-rearing periods.

Carbon markets were highlighted in the preceding section as related to mitigation that may result in changes to wildlife habitat. Building upon this notion, emerging carbon markets also may serve as an incentive program for promoting land management for wildlife habitat. Diaz et al. (2009) described compliance and voluntary markets as the two existing categories of carbon markets in the United States. Compliance markets function through regulatory organizations that limit the number of offset allowances (for the polluter) and credits (for offseting pollution activities) that are available throughout a marketplace. Commonly called cap-and-trade systems, such a system was established by the U.S. Environmental Protection Agency to reduce pollution as part of the Acid Rain Program (Diaz et al. 2009). Essentially, pollution allowances not used by one organization may be sold to another organization that has not met emission reduction requirements. Voluntary markets, however, function as an altruistic mechanism for which participation is influenced by responsibility to society. As noted by Diaz et al. (2009), the two current types of voluntary carbon markets are the Chicago Climate Exchange and over-the-counter transactions. The Chicago Climate Exchange involves trading emission allowances and credits among member companies, whereas over-the-counter transactions are based on privately established contracts (Diaz et al. 2009). In terms of actual transactions, buyers of ecosystem services purchase carbon credits on the Chicago Climate Exchange to offset activities that emit greenhouse gases. In turn, the Chicago Climate Exchange buys credits from projects that offset the accumulation of greenhouse gases in the atmosphere. Many existing forestry programs that can enhance wildlife habitat also offset greenhouse gas emissions, and some are already generating income for forest-land owners. However, the total dollar amount of traded projects has been relatively limited, and carbon prices on the exchange in recent times have fallen below $1 per ton (see chapter 2).

Regarding the future, there is substantial interest in identifying approaches and policies to minimize impacts of climate change on wildlife. Mawdsley et al. (2009) suggested strategies for wildlife management and biodiversity conservation as related to anticipated impacts of climate change. Strategies described include, among others, increasing size of protected areas, designing new natural areas and restoration sites, protecting land to increase connectivity, managing for ecosystem processes, reducing deforestation (e.g., Alig et al. 2010a), and enhancing the landscape to support movements by many species. Incorporating anticipated impacts of climate change into any land management plan and programs likely will be costly and difficult to implement across all involved organizations (Mawdsley et al. 2009). Any strategy is likely to require cooperation with private landowners. For example, increasing the size of a protected area likely will need to account for private land holdings beyond the protected area boundary. At the same time, policies that specifically target wildlife habitat may not maximize other benefits (Nelson et al. 2008). At this time, there is a need to develop knowledge about which strategies may be feasible, ethical, and potentially successful.

Linkages of Climate Change Model Projections and Impacts From Global to Local Scales

Conceptually, the process of projecting adaptation to and mitigation in response to climate change impacts on wildlife requires multiple levels of analysis, and consideration of wildlife habitat at a unit of observation (or scale) appropriate for ecology of individual species and meaningful to a policy analysis. Thinking hierarchically and within a systems analysis perspective, a constellation of individual assessment tools may be used, such as global climate models, regional climate models, vegetation (e.g., forest inventory projection) and land cover and land use models, species-level bioclimatic envelope models, and site-specific analyses that may consider local abiotic and biotic characteristics (Hannah et al. 2002, Sulzman et al. 1995). All of these components also need to be able to model key feedbacks, both positive and negative. Because of constantly evolving technology, rapid information exchange, and broadening intellectual resources, the process of climate change research and projection modeling is extremely dynamic. We provide a brief review of assessment tools, but because of limited understanding about interactions between possible future adaptation and mitigation activities, that topic is left for future research and is not covered here.

The constellation of assessment tools spans from global to local. Broadly at the global and limited regional levels, general circulation models (GCMs) simulate atmospheric characteristics (e.g., radiation, precipitation) and interactions between the atmosphere and land surface characteristics that affect climate (e.g., oceans, land topography) (Sulzman et al. 1995). General circulation models are used by climate policy authorities (IPCC 2007), and are updated continuously with development of new research tools (Pope et al. 2002). Variation in prediction exists among GCMs, as each differs based on parameterization characteristics such as resolution of analysis (Murphy et al. 2004). The GCMs are a typical starting place for projecting climate change effects on biodiversity because they are the only tools that can consider broad complex atmospheric processes and variability (Hannah et al. 2002, Murphy et al. 2004).

To evaluate effects of climate change on biodiversity, GCMs often are used in tandem with land cover and land use models (Sulzman et al. 1995; see Peng and Wen (2006) for an overview of major types of forest simulation models). Ravenscroft et al. (2010) used the forest ecosystem model LANDIS-II to evaluate projected interactions of forest management and climate change in northern Minnesota. Results suggested that climate change may result in an overall decreased forest-tree diversity, and replacement of current forest composition by other dominant tree species. As another example, Cramer et al. (2001) used an ocean-atmospheric model coupled with a combination of six dynamic global vegetation models (DGVMs) to estimate impacts of climate change on vegetation carbon and water exchange and carbon storage. From this study, results differed among DGVMs, but all six models projected similar trends in decreasing carbon storage (sink) after 2050. The authors suggested the projected decrease in carbon storage to be partially a result of respiration among heterotrophs (species that do not photosynthesize their food), projected variation and reductions in precipitation, and deforestation (Cramer et al. 2001).

In addition to DGVMs, investigators also have used biogeochemistry and biogeographic models that simulate the gain, loss, and internal cycling of carbon, nutrients, and water. Biochemistry models project climate impacts on changes in temperature, precipitation, soil moisture, atmospheric carbon dioxide, and other climate-related factors, which can be examined for their influence on such processes as ecosystem productivity and carbon storage. Biogeographic models examine the influence of climate on the geographic distribution of plant species or plant types such as trees, grasses, and shrubs. For example, Alig et al. (2002) used scenarios representing combinations of climate projections from two GCMs (Canadian and Hadley), which in turn were fed into two ecological process models that are classified as biogeochemical models (Century and Terrestrial Ecosystem Models) (Irland et al. 2001). The latter models simulate the impacts of climate on forest productivity. Estimates of changes in forest productivity were fed into the FASOM-GHG[1] economic optimization model (Alig et al. 2002). The modeling includes forest land converted to urban/developed (Alig et al. 2010b) and agricultural uses, as well as afforested land, with responses to different policy incentives such as carbon prices (Alig et al. 2010a). The authors presented a wide variety of outputs, including changes in forest age classes, forest types, vegetation carbon, and other attributes that may affect wildlife habitat.

Climate change is expected to influence multiple sectors of the global economy. Agarwal et al. (2002) described methods for integrating land use change models into evaluations of impacts of climate change. As related to the forest sector, climate change effects on forests may consequently impact the forest product market (Perez-Garcia et al. 2002a, 2002b). Sohngen et al. (2001) projected that timber production will increase in low- and mid-latitude forests as a result of planting of short-rotation plantations, whereas higher latitude areas may experience a decrease in production based on planting of relatively long-rotation species. Irland et al.

---

[1] Forest and Agricultural Sector Optimization Model–Greenhouse Gas.

Climate change may affect both market and nonmarket segments of the forest sector.

(2001) suggested that climate change may affect not only forest products, but also nonmarket sectors of the forest economy such as forest-based outdoor recreation. The authors suggested that particular activities that rely on winter precipitation, such as downhill skiing, may be affected by climate change, but potential impacts are uncertain.

Returning the focus to individual species and communities, another type of model used to evaluate impacts of climate change on wildlife are climate (or environmental or bioclimatic) envelope models. These models consider the distribution of a species as under an "envelope" of particular environmental conditions, and then evaluate how the location of the envelope may change under different climate change scenarios. Lawler et al. (2009) used a climate envelope approach coupled with 30 scenarios developed by use of an atmosphere-ocean general circulation model (AOGCM) to project changes in geographic ranges of 2,954 vertebrate species (amphibians, birds, and mammals) in the Western Hemisphere. The authors reported that climate change projections suggest varying amounts of species losses and gains across the Western Hemisphere. Ultimately, the same authors projected relatively large changes in local species communities (fig. 1-4). From the same study, it is expected that the greatest changes would be among amphibians and, at higher latitudes and in the Andes Mountains and Central America (including

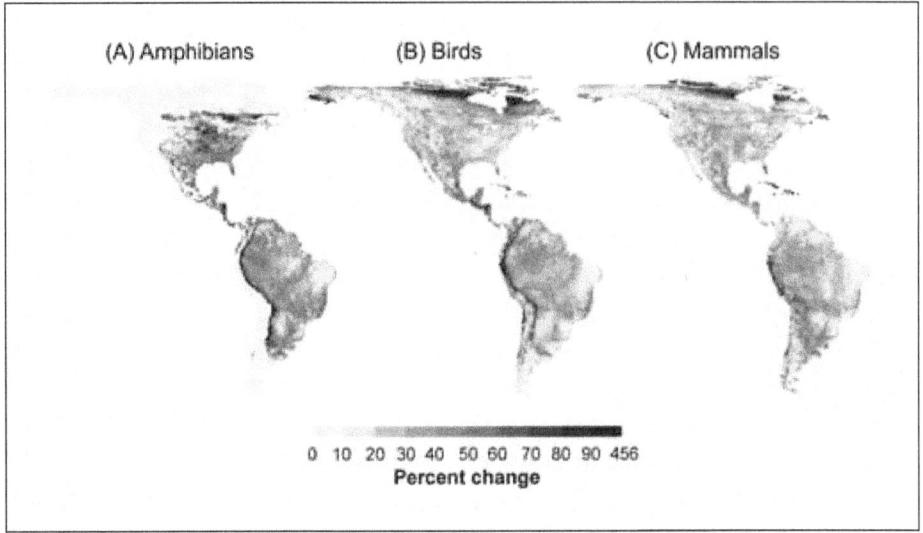

Figure 1-4—Predictions of species composition turnover for (A) amphibians, (B) birds, and (C) mammals as a result of climate change. Such turnover may result in drastic differences in composition of local wildlife communities. Adapted from Lawler et al. 2009.

Mexico), all vertebrate groups (Lawler et al. 2009). For climate envelope models, some research suggests that occurrence of rare species may be more sensitive to stochastic events than more common species (Green et al. 2008), that spatial scale of analysis is of considerable importance (Pearson and Dawson 2003), and that results may differ based on species used in the analysis (Araujo et al. 2009, Beale et al. 2008). However, such models may be useful tools for policymakers for evaluating environmental indicators of interest.

Several other modeling techniques may be used to link wildlife habitat information to climate change projections, and are mentioned only briefly here. For example, ecological niche models incorporate a computer algorithm and spatial data layers to estimate potential distribution of species. Peterson (2003) coupled ecological niche and climate model scenarios to evaluate the role of topography in affecting distribution of bird species in the western and central North America. As another example, Maxent models are used to integrate land cover and climate-related variables to assess habitat suitability (Stabach et al. 2009). In addition, simulation models and tools that may be parameterized with wildlife life history information are used to evaluate effects of landscape change on individual species (Lookingbill et al. 2010, Schumaker et al. 2004).

It is important to note, however, that any modeling approach to address global climate change has its limitations, and projections should be interpreted with caution. Even with efforts to obtain the best analytical results, to state implications of model projections across scales (global to local) requires the user to make assumptions about how results of one model may connect to or integrate well with the parameters of another model. As noted earlier in this section, climate model projections can differ widely based on parameterization methods (Murphy et al. 2004). It is also difficult to evaluate performance of climate models, as future environmental characteristics are unknown (Araújo et al. 2005, Thullier 2007). In fact, some researchers suggest a coarse-filter approach and other spatial analysis tools rather than coupling projection models with high uncertainty (Beier and Brost 2010). Additionally, because vegetation land cover and particular microhabitat features are important components of wildlife habitat, projecting climate impacts on vegetation communities is an important linkage for assessment of climate impacts on wildlife, and often is not considered when coupling GCMs and species range maps to make projections (Botkin et al. 2007). Although a critique of different modeling tools is beyond the scope of this paper, Botkin et al. (2007) provided a comprehensive comparison of different methods to project effects of climate change on biodiversity.

Thinking Toward the Future

Climate change is a complex and far-reaching issue, both spatially and temporally, and involves a relatively large amount of uncertainty even beyond the typical human inability to predict the future. Both adaptation and mitigation actions in the future are unknown, as is their interaction over time and space. An adaptive process will facilitate incorporation of research results into management, as well as ongoing consideration of land-use changes, forest management, wildlife management, mitigation alternatives, and society's willingness to undertake actions to formulate policy alternatives and confront other components of the global climate issue. Needs for additional knowledge focus, for example, on the following areas:

- For individual species and populations, we have the most empirical knowledge about range shifts and migration patterns; less is known about nonmigratory species, particularly over the long term.
- Effects of range shifts on competition between new species and species already within an area, and adaptive behaviors of species that are successfully adjusting to climate change impacts.
- Small-scale studies to verify and validate large-scale patterns and modeling projections (Root and Schneider 2006).
- Assessment of multiple stressors such as direct and additive impacts of concurrent land use and climate change (Inkley et al. 2004), and potential climate-related effects on landscape characteristics such as habitat connectivity. Few studies have incorporated empirical land use and climate

data concurrently (see McAlpine et al. 2009), particularly for building of (see Knowlton and Graham 2010) and application to (see McRae et al. 2008) models designed to make projections for individual plant and animal species.

- Indirect effects of multiple climate-related abiotic stressors, such as the aforementioned cascading effects of precipitation, snowmelt, erosion, and sediment loads in rivers on trout habitat and spawning (Scheurer et al. 2009).

- Integrating projected climate-related impacts on vegetation communities to make stronger linkages between GCMs and wildlife information (Botkin et al. 2007).

- Assessment of vulnerability of species in habitats in proximity to development, and evaluation of adaptation strategies for and adaptive management of climate change impacts on wildlife and wildlife habitat. Mawdsley et al. (2009) suggested numerous potential adaptation strategies as a foundation, but more research is needed to determine the feasibility, ethics, and potential success of such strategies.

Climate change is a complex and uncertain issue. It affects and is affected by both natural and social systems at multiple scales ranging from global to local. In one perspective, more weight may be warranted for short-term research that may support and inform management and policy decisions regarding both adaptation and mitigation. For the case of wildlife, for which potential ability to adapt to changing conditions may be limited, there may not be time to pursue long-term research while avoiding varying levels of extinction. Rather, research results may be valuable as information to use within broader and immediate adaptive management processes dealing with climate change, while also maintaining natural resource management and societal objectives.

English Equivalents

When you know:	Multiply by:	To get:
Meters (m)	3.28	Feet
Kilometers (km)	.621	Miles
Hectares (ha)	2.47	Acres
Degrees Celsius (°C)	1.8 °C + 32	

Literature Cited

Agarwal, C.; Green, G.M.; Grove, J.M.; Evans, T.P.; Schweik, C.M. 2002. A review and assessment of land-use change models. Gen. Tech. Rep. NE-297. Newton Square, PA: U.S. Department of Agriculture, Forest Service, Northeastern Research Station. 61 p.

Alig, R.J. 2010. Economic modeling of effects of climate change on the forest sector and mitigation options: a compendium of briefing papers. Gen. Tech. Rep. PNW-GTR-833. Portland, OR: U.S. Department of Agriculture, Forest Service, Pacific Northwest Research Station. 169 p.

Alig, R.J.; Adams, D.M.; McCarl, B. 2002. Projecting impacts of global change on the U.S. forest and agricultural sectors and carbon budgets. Forest Ecology and Management. 169: 3–14.

Alig, R.J.; Latta, G.; Adams, D.M.; McCarl, B. 2010a. Mitigating greenhouse gases: the importance of land base interactions between forests, agriculture, and residential development in the face of changes in bioenergy and carbon prices. Forest Policy and Economics. 12: 67–75.

Alig, R.J.; Plantinga, A.; Haim, D.; Todd, M. 2010b. Area changes in U.S. forests and other major land uses, 1982–2002, with projections to 2062. Gen. Tech. Rep. PNW-GTR-815. Portland, OR: U.S. Department of Agriculture, Forest Service, Pacific Northwest Research Station. 102 p.

Allen, A.; Chapman, D. 2001. Impacts of afforestation on groundwater resources and quality. Hydrogeology Journal. 9: 390–400.

Allen, C.D.; Breshears, D.D. 1998. Drought-induced shift of a forest-woodland ecotone: rapid landscape response to climate variation. Proceedings of the National Academy of Sciences. 95: 14839–14842.

Araújo, M.B.; Pearson, R.G.; Thuiller, W.; Erhard, M. 2005. Validation of species-climate impact models under climate change. Global Change Biology. 11: 1504–1513.

Araújo, M.B.; Thuiller, W.; Yoccoz, N.G. 2009. Reopening the climate envelope reveals macroscale associations with climate in European birds. Proceedings of the National Academy of Sciences. 106: E45–E46.

The Audubon Society. 2010. Birds and climate change: on the move. http://birdsandclimate.audubon.org/. (5 May 2010).

Backlund, P.; Janetos, A.; Schimel, D. 2008. The effects of climate change on agriculture, land resources, water resources, and biodiversity in the United States. Synthesis and Assessment Product 4.3. Report by the U.S. Climate Change Science Program and the Subcommittee on Global Change Research. Washington, DC: Department of Commerce and Department of Energy. http://www.climatescience.gov/Library/sap/sap4-3/final-report/default.htm. (16 June 2010).

Beale, C.M.; Lennon, J.L.; Gimona, A. 2008. Opening the climate envelope reveals no macroscale associations with climate in European birds. Proceedings of the National Academy of Sciences. 105: 14908–14912.

Beebee, T.J.C.; Griffiths, R.A. 2005. The amphibian decline crisis: a watershed for conservation biology. Biological Conservation. 125: 271–285.

Beever, E.A.; Brussard, P.F.; Berger, J. 2003. Patterns of apparent extirpation among isolated populations of pikas (*Ochotona princeps*) in the Great Basin. Journal of Mammalogy. 84: 37–54.

Beier, P.; Brost, B. 2010. Use of land facets to plan for climate change: conserving the arenas, not the actors. Conservation Biology. 24: 701–710.

Berg, E.E.; Henry, J.D.; Fastie, C.L.; De Volder, A.D.; Matsuoka, S.M. 2006. Spruce beetle outbreaks on the Kenai Peninsula, Alaska, and Kluane National Park and Reserve, Yukon Territory: relationship to summer temperatures and regional differences in disturbance regimes. Forest Ecology and Management. 227: 219–232.

Berger, J. 2004. The last mile: how to sustain long-distance migration in mammals. Conservation Biology. 18: 320–331.

Berkes, F.; Jolly, D. 2001. Adapting to climate change: social-ecological resilience in a Canadian western arctic community. Conservation Ecology. 5: 18. http://www.ecologyandsociety.org/vol5/iss2/art18/. (23 December 2009).

Bies, L. 2006. The biofuels explosion: Is green energy good for wildlife? The Wildlife Society Bulletin. 34: 1203–1205.

Blaustein, A.R.; Belden, L.K.; Olson, D.H.; Green, D.M.; Root, T.L.; Kiesecker, J.M. 2001. Amphibian breeding and climate change. Conservation Biology. 15: 1804–1809.

Blaustein, A.R.; Romansic, J.R.; Kiesecker, J.M.; Hatch, A.C. 2003. Ultraviolet radiation, toxic chemicals, and amphibian population declines. Diversity and Distributions. 8: 123–140.

Blaustein, A.R.; Wake, D.B. 1995. The puzzle of declining amphibian populations. Scientific American. 272: 52–57.

Botkin, D.B.; Saxe, H.; Araújo, M.B.; Betts, R.; Bradshaw, R.H.W.; Cedhagen, T.; Chesson, P.; Dawson, T.P.; Etterson, J.R.; Faith, D.P.; Ferrier, S.; Guisan, A.; Hansen, A.S.; Hilbert, D.W.; Loehle, C.; Margules, C.; New, M.; Sobel, M.J.; Stockwell, D.R.B. 2007. Forecasting the effects of global warming on biodiversity. BioScience. 57: 227–236.

Bradley, N.L.; Leopold, A.C.; Ross, J.; Huffaker, W. 1999: Phenological changes reflect climate change in Wisconsin. Proceedings of the National Academy of Sciences. 96: 9701–9704.

Carey, C.; Alexander, M.A. 2003. Climate change and amphibian declines: Is there a link? Diversity and Distributions. 9: 111–121.

Chen, I-C.; Shiu, H-J.; Benedick, S.; Holloway, J.D.; Chey, V.K.; Barlow, H.S.; Hill, J.K.; Thomas, C.D. 2009. Elevation increases in moth assemblages over 42 years on a tropical mountain. Proceedings of the National Academy of Sciences. 106: 1479–1483.

Cramer, W.; Bondeau, A.; Woodward, F.I.; Prentice, I.C.; Betts, R.A.; Brovkin, V.; Cox, P.M.; Fisher, V.; Foley, J.A.; Friend, A.D.; Kucharik, C.; Lomas, M.R.; Ramankutty, N.; Sitch, S.; Smith, B.; White, A.; Young-Molling, C. 2001. Global response of terrestrial ecosystem structure and function to CO_2 and climate change: results from six dynamic global vegetation models. Global Change Biology. 7: 357–373.

Crozier, L. 2003. Winter warming facilitates range expansion: cold tolerance of the butterfly *Atalopedes campestris*. Oecologia. 135: 648–656.

Crozier, L. 2004. Warmer winters drive butterfly range expansion by increasing survivorship. Ecology. 85: 231–241.

Dale, V.H.; Joyce, L.A.; McNulty, S.; Neilson, R.P. 2000. The interplay between climate change, forests, and disturbances. The Science of the Total Environment. 262: 201–204.

Descimon, H.; Bachelard, P.; Boitier, E.; Pierrat, V. 2005. Decline and extinction of *Parnassius apollo* populations in France—continued. Studies on the ecology and conservation of butterflies in Europe. In: Kühn, E.; Feldmann, R.; Thomas, J.A; Settel, J., eds. Concepts and case studies. Sofia, Bulgaria: Pensoft Publishers: 114–115. Vol. 1.

Diaz, D.D.; Charnley, S.; Gosnell, H. 2009. Engaging western landowners in climate change mitigation: a guide to carbon-oriented forest and range management and carbon market opportunities. Gen. Tech. Rep. PNW-GTR-801. Portland, OR: U.S. Department of Agriculture, Forest Service, Pacific Northwest Research Station. 81 p.

Fargione, J.E.; Cooper, T.R.; Flaspohler, D.J.; Hill, J.; Lehman, C.; McCoy, T.; McLeod, S.; Nelson, E.J.; Oberhauser, K.S.; Tilman, D. 2009. Bioenergy and wildlife: threats and opportunities for grassland conservation. BioScience. 59: 767–777.

Feng, H.; Kurkalova, L.A.; Kling, C.L.; Gassman, P.W. 2006. Environmental conservation in agriculture: land retirement vs. changing practices on working land. Journal of Environmental Economics and Management. 52: 600–614.

Feng, H.; Kurkalova, L.A.; Kling, C.L.; Gassman, P.W. 2007. Transfers and environmental co-benefits of carbon sequestration in agricultural soils: retiring agricultural land in the Upper Mississippi River Basin. Climate Change. 80: 91–107.

Flannigan, M.D.; Stocks, B.J.; Wotton, B.M. 2000. Climate change and forest fires. The Science of the Total Environment. 262: 221–229.

Gaston, K.J. 2003. The structure and dynamics of geographic ranges. Oxford, United Kingdom: Oxford University Press. 280 p.

Gibbs, J.P.; Breisch, A.R. 2001. Climate warming and calling phenology of frogs near Ithaca, New York, 1900–1999. Conservation Biology. 15: 1175–1178.

Green, R.E.; Collingham, Y.C.; Willis, S.G.; Gregory, R.D.; Smith, K.W.; Huntley, B. 2008. Performance of climate envelope models in retrodicting recent changes in bird population size from observed climatic change. Biology Letters. 4: 599–602.

Griffith, B.; Douglas, D.C.; Walsh, N.E.; Young, D.D.; McCabe, T.R.; Russell, D.E.; White, R.G.; Cameron, R.D.; Whitten, K.R. 2002. The Porcupine caribou herd. In: Douglas, D.C.; Reynolds, P.E.; Rhode, E.B., eds. Arctic Refuge coastal plain terrestrial wildlife research summaries. Biological Science Rep. USGS/BRD BSR-2002-0001. Reston, VA: U.S. Geological Survey: 8–37.

Gwinner, E. 1996. Circannual clocks in avian reproduction and migration. Ibis. 138: 47–63.

Hannah, L.; Midgley, G.F.; Millar, D. 2002. Climate change-integrated conservation strategies. Global Ecology and Biogeography. 11: 485–495.

Harcombe, P.A.; Hall, R.B.W.; Glitzenstein, J.S.; Cook, E.S.; Krusic, P.; Fulton, M.; Streng, D.R. 1998. Sensitivity of Gulf Coast forests to climate change. In: Guntenspergen, G.R.; Vairin, B.A., eds. Vulnerability of coastal wetlands in the Southeastern United States: climate change research results, 1992–97. Biological Science Report USGS/BRD/BSR 1998-0002. Lafayette, LA: U.S. Geological Survey: 45–66.

Harris, G.; Thirgood, S.; Hopcraft, J.G.C.; Cromsigt, J.P.G.M.; Berger, J. 2009. Global decline in aggregated migrations of large terrestrial mammals. Endangered Species Research. 7: 55–76.

Harsch, M.A.; Hulme, P.E.; McGlone, M.S.; Duncan, R.P. 2009. Are treelines advancing? A global meta-analysis of treeline response to climate warming. Ecology Letters. 12: 1040–1049.

Herkert, J.R. 2007. Evidence for a recent Henslow's sparrow population increase in Illinois. Journal of Wildlife Management. 71: 1229–1233.

Hughes, L. 2000. Biological consequences of global warming: Is the signal already apparent? Trends in Ecology and Evolution. 15: 56–61.

Inkley, D.B.; Anderson, M.G.; Blaustein, A.R.; Burkett, V.R.; Felzer, B.; Griffith, B.; Price, J.; Root, T.L. 2004. Global climate change and wildlife in North America. Wildlife Society Tech. Review 04-2. Bethesda, MD: The Wildlife Society. 26 p.

Intergovernmental Panel on Climate Change [IPCC]. 2007. Synthesis report. http://www.ipcc.ch/pdf/assessment-report/ar4/syr/ar4_syr.pdf. (30 December 2009).

Irland, L.C.; Adams, D.A.; Alig, R.J.; Betz, C.J.; Chen, C.C.; Hutchins, M.; McCarl, B.A.; Skog, K.; Sohngen, B.L. 2001. Assessing socioeconomic impacts of climate change on U.S. forests, wood-product markets, and forest recreation. BioScience. 51: 753–764.

Issac, J.L. 2009. Effects of climate change on life history: implications for extinction risk in mammals. Endangered Species Research. 7: 115–123.

Iverson, L.R.; Prasad, A.M.; Matthews, S.N.; Peters, M. 2008. Estimating potential habitat for 134 Eastern U.S. tree species under six climate scenarios. Forest Ecology and Management. 254: 390–406.

Johnson, R. 1995. Supplemental sources of income for southern timberland owners. Journal of Forestry. 93: 22–24.

Kausrud, K.L.; Mysterud, A.; Steen, H.; Vik, J.O.; Østbye, E.; Cazelles, B.; Framstad, E.; Eikeset, A.M.; Mysterud, I.; Solhøy, T.; Stenseth, N.C. 2008. Linking climate change to lemming cycles. Nature. 456: 93–98.

Kiesecker, J.M.; Blaustein, A.R.; Belden, L.K. 2001. Complex causes of amphibian population declines. Nature. 410: 681–684.

Knops, J.M.H.; Tilman, D.; Haddad, N.M.; Naeem, S.; Mitchell, C.E.; Haarstad, J.; Ritchie, M.E.; Howe, K.M.; Reich, P.B.; Siemann, E.; Groth, J. 1999. Effects of plant species richness on invasion dynamics, disease outbreaks, insect abundances and diversity. Ecology Letters. 2: 286–293.

Knowlton, J.L.; Graham, C.H. 2010. Using behavioral landscape ecology to predict species' responses to land-use and climate change. Biological Conservation. 143: 1342–1354.

Körner, C. 2007. The use of "altitude" in ecological research. Trends in Ecology and Evolution. 22: 569–574.

Körner, C.; Paulsen, J. 2004. A world-wide study of high altitude treeline temperatures. Journal of Biogeography. 31: 713–732.

Kurkalova, L.; Kling, C.; Zhao, J. 2006. Green subsidies in agriculture: estimating the adoption costs of conservation tillage from observed behavior. Canadian Journal of Agricultural Economics. 54: 247–267.

Kurz, W.A.; Dymond, C.C.; Stinson, G.; Ramley, G.J.; Neilson, E.T.; Carroll, A.L.; Ebata, T.; Safranyik, L. 2008. Mountain pine beetle and forest carbon feedback to climate change. Nature. 452: 987–990.

Kyriacou, C.P. 2010. Unraveling traveling. Science. 325: 1629–1630.

Laliberté, E.; Wells, J.A.; DeClerck, F.; Metcalfe, D.J.; Catterall, C.P.; Queiroz, C.; Aubin, I.; Bonser, S.P.; Ding, Y.; Fraterrigo, J.M.; McNamara, S.; Morgan, J.W.; Sánchez Merlos, D.; Vesk, P.A.; Mayfield, M.M. 2010. Land-use intensification reduces functional redundancy and response diversity in plant communities. Ecological Letters. 13: 76–86.

Lawler, J.J.; Shafer, S.L.; White, D.; Kareiva, P.; Maurer, E.P.; Blaustein, A.R.; Bartlein, P.J. 2009. Projected climate-induced faunal change in the western hemisphere. Ecology. 90: 588–597.

Ledneva, A.; Miller-Rushing, J.A.; Primack, R.B.; Imbres, C. 2004. Climate change as reflected in a naturalist's diary, Middleborough, Massachusetts. Wilson Bulletin. 116: 224–231.

Logan, J.A.; Powell, J.A. 2001. Ghost forests, global warming, and the mountain pine beetle (*Coleoptera: Scolytidae*). American Entomologist. 47: 160–173.

Lookingbill, T.R.; Gardner, R.H.; Ferrari, J.R.; Keller, C.E. 2010. Combining a dispersal model with network theory to assess habitat connectivity. Ecological Applications. 20: 427–441.

Matthews, S.; O'Connor, R.; Plantinga, A.J. 2002. Quantifying the impacts on biodiversity of policies for carbon sequestration in forests. Ecological Economics. 40: 71–87.

Mawdsley, J.R.; O'Malley, R.; Ojima, D.S. 2009. A review of climate-change adaptation strategies for wildlife management and biodiversity conservation. Conservation Biology. 23: 1080–1089.

McAlpine, C.A.; Syktus, J.; Ryan, J.G.; Deo, R.C.; McKeon, G.M.; McGowan, H.A.; Phinn, S.R. 2009. A continent under stress: interactions, feedbacks and risks associated with impact of modified land cover on Australia's climate. Global Change Biology. 15: 2206–2223.

McRae, B.H.; Schumaker, N.H.; McKane, R.B.; Busing, R.T.; Solomon, A.M.; Burdick, C.A. 2008. A multi-model framework for simulating wildlife population response to land-use and climate change. Ecological Modelling. 219: 77–91.

Menzel, A.; Sparks, T.H.; Estrella, N.; Koch, E.; Aasa, A.; Ahas, R.; Alm-Kübler, K.; Bissolli, P.; Braslavská, O.; Briede, A.; Chmielewski, F.M.; Crepinsek, Z.; Curnel, Y.; Dahl, Å.; Defila, C.; Donnelly, A.; Filella, Y.; Jatczak, K.; Måge, F.; Mestre, A.; Nordli, Ø.; Peñuelas, J.; Pirinen, P.; Remišová, V.; Scheifinger, H.; Striz, M.; Susnik, A.; van Vliet, A.J.H.; Wielgolaski, F.E.; Zach, S.; Zust, A. 2006. European phonological response to climate changes matches the warming pattern. Global Change Biology. 12: 1969–1976.

Millennium Ecosystem Assessment [MEA]. 2005. Ecosystems and human well-being: biodiversity synthesis. Washington, DC: World Resources Institute. 86 p.

Moritz, C.; Patton, J.L.; Conroy, C.J.; Parra, J.L.; White, G.C.; Bessinger, S.R. 2008. Impact of a century of climate change on small-mammal communities in Yosemite National Park, USA. Science. 322: 261–264.

Murphy, J.M.; Sexton, D.M.H.; Barnett, D.N.; Jones, G.S.; Webb, M.J.; Collins, M.; Stainforth, D.A. 2004. Quantification of modeling uncertainties in a large ensemble of climate change simulations. Nature. 430: 768–772.

Nelson, E.; Polasky, S.; Lewis, D.J.; Plantinga, A.J.; Lonsdorf, E.; White, D.; Bael, D.; Lawler, J.J. 2008. Efficiency of incentives to jointly increase carbon sequestration and species conservation on a landscape. Proceedings of the National Academy of Sciences. 105: 9471–9476.

Niemuth, N.D.; Quamen, F.R.; Naugle, D.E.; Reynolds, R.E.; Estey, M.E.; Shaffer, T.L. 2007. Benefits of the Conservation Reserve Program to grassland bird populations in the Prairie Pothole Region of North Dakota and South Dakota. Report prepared for the USDA Farm Service Agency. Reimbursable Fund Agreement OS-IA-04000000-N34. On file with: Anita T. Morzillo, Department of Forest Ecosystems and Society, College of Forestry, Oregon State University, Corvallis, OR 97331.

North American Bird Conservation Initiative, U.S. Committee [NABCI]. 2009. The state of the birds, United States of American, 2009. Washington, DC: U.S. Department of the Interior. 36 p.

North American Bird Conservation Initiative, U.S. Committee [NABCI]. 2010. The state of the birds, United States of America, 2010. Washington, DC: U.S. Department of the Interior. 32 p.

Palmer, T.M.; Stanton, M.L.; Young, T.P.; Goheen, J.R.; Pringle, R.M.; Karban, R. 2008. Breakdown of an ant-plant mutualism follows the loss of large herbivores from an African savanna. Science. 319: 192–195.

Parmesan, C. 1996. Climate and species range. Nature. 382: 765–766.

Parmesan, C. 2006. Ecological and evolutionary responses to recent climate change. Annual Review of Ecology, Evolution, and Systematics. 37: 637–669.

Parmesan, C.; Galbraith, H. 2004. Observed impacts of global climate change in the U.S. A report for the Pew Center on Global Climate Change. www.pewclimate.org. (5 May 2010).

Parmesan, C.; Ryrholm, N.; Stefanescu, C.; Hill, J.K.; Thomas, C.D.; Descimon, H.; Huntley, B.; Kaila, L.; Kullberg, J.; Tammaru, T.; Tennent, W.J.; Thomas, J.A.; Warren, M. 1999. Poleward shifts in geographical ranges of butterfly species associated with regional warming. Nature. 399: 579–583.

Parmesan, C.; Yohe, G. 2003. A globally coherent fingerprint of climate change impacts across natural systems. Nature. 421: 37–42.

Pearson, R.G.; Dawson, T.P. 2003. Predicting the impacts of climate change on the distribution of species: Are bioclimate envelope models useful? Global Ecology and Biogeography. 12: 361–371.

Peng, C.; Wen, X. 2006. Forest simulation models. In: Shao, G.; Reynolds, K.M., eds. Computer applications in sustainable forest management. New York: Springer: 101–125.

Perez-Garcia, J.; Joyce, L.A.; McGuire, A.D. 2002a. Temporal uncertainties of integrated ecological/economic assessments at the global and regional scales. Forest Ecology and Management. 162: 105–115.

Perez-Garcia, J.; Joyce, L.A.; McGuire, A.D.; Xiao, X. 2002b. Impacts of climate change on the global forest sector. Climate Change. 54: 439–461.

Peterson, A.T. 2003. Projected climate change effects on Rocky Mountain and Great Plains birds: generalities of biodiversity consequences. Global Change Biology. 9: 647–655.

Plantinga, A.J. 1996. The effect of agricultural policies on land use and environmental quality. American Journal of Agricultural Economics. 78: 1082–1091.

Plantinga, A.J.; Wu, J. 2003. Co-benefits from carbon sequestration in forests: evaluating reductions in agricultural externalities from an afforestation policy in Wisconsin. Land Economics. 79: 74–85.

Pope, V.D.; Gallani, M.L.; Rowntree, P.R.; Stratton, R.A. 2002. The impact of new physical parameterizations in the Hadley Centre climate model: HadAM3. Climate Dynamics. 16: 123–146.

Popescu, V.D.; Gibbs, J.P. 2009. Interactions between climate, beaver activity, and pond occupancy by the cold-adapted mink frog in New York State, USA. Biological Conservation. 142: 2059–2068.

Pounds, J.A.; Bustamante, M.R.; Coloma, L.A.; Consuegra, J.A.; Fogden, M.P.L.; Foster, P.N.; La Marca, E.; Masters, K.L.; Merino-Viteri, A.; Puschendorf, R.; Ron, S.R.; Sánchez-Azofeifa, G.A.; Still, C.J.; Young, B.E. 2006. Widespread amphibian extinctions from epidemic disease driven by global warming. Nature. 439: 161–167.

Pounds, J.A.; Crump, M.L. 1994. Amphibian declines and climate disturbance: the case of the golden toad and the harlequin frog. Conservation Biology. 8: 72–85.

Pounds, J.A.; Fogden, M.P.L.; Campbell, J.H. 1999. Biological responses to climate change on a tropical mountain. Nature. 398: 611–615.

Price, J.; Glick, P. 2002. The birdwatcher's guide to global warming. National Wildlife Federation and American Bird Conservancy. http://www.abcbirds.org/newsandreports/birdguide.html. (24 November 2009).

Price, J.T.; Root, T.L. 2005. Potential impacts of climate change on Neotropical migrants: management implications. Gen. Tech. Rep. PSW-GTR-191. Albany, CA: U.S. Department of Agriculture, Forest Service, Pacific Southwest Research Station. 1294 p.

Ravenscroft, C.; Scheller, R.M.; Mladenoff, D.J.; White, M.A. 2010. Forest restoration in a mixed-ownership landscape under climate change. Ecological Applications. 20: 327–346.

Robinson, R.A.; Crick, H.Q.P.; Learmonth, J.A.; Maclean, I.M.D.; Thomas, C.D.; Bairlein, F.; Forchhammer, M.C.; Francis, C.M.; Gill, J.A.; Godley, B.J.; Harwood, J.; Hays, G.C.; Huntley, B.; Hutson, A.M.; Pierce, G.J.; Rehfisch, M.M.; Sims, D.W.; Santos, M.B.; Sparks, T.H.; Stroud, D.A.; Visser, M.E. 2009. Travelling through a warming world: climate change and migratory species. Endangered Species Research. 7: 87–99.

Robles, M.D.; Flather, C.H.; Stein, S.M.; Nelson, M.D.; Cutko, A. 2008. The geography of private forests that support at-risk species in the conterminous United States. Frontiers in Ecology and the Environment 6. http://www. esajournals.org/doi/abs/10.1890/070106. (30 August 2009).

Rohr, J.R.; Raffel, T.R.; Romansic, J.M.; McCallum, H.; Hudson, P.J. 2008. Evaluating the links between climate, disease spread, and amphibian declines. Proceedings of the National Academy of Sciences. 105: 17436–17441.

Root, T.L.; Price, J.T.; Hall, K.R.; Schneider, S.H.; Rosenzweig, C.; Pounds, J.A. 2003. Fingerprints of global warming on wild animals and plants. Nature. 421: 57–60.

Root, T.L.; Schneider, S.H. 2006. Conservation and climate change: the challenges ahead. Conservation Biology. 20: 706–708.

Sawyer, H.; Lindzey, F.; McWhirter, D. 2007. Mule deer and pronghorn migration in western Wyoming. Wildlife Society Bulletin. 33: 1266–1273.

Sawyer, H.; Lindzey, F.; McWhirter, D.; Andrews, K. 2002. Potential effects of oil and gas development on mule deer and pronghorn populations in western Wyoming. Transactions of the 67th North American Wildlife and Natural Resources Conference. Washington, DC: Wildlife Management Institute. 67: 350–365.

Scheurer, K.; Alewell, C.; Bänninger, D.; Burkhardt-Holm, P. 2009. Climate and land-use changes affecting river sediment and brown trout in alpine countries—a review. Environmental Science and Pollution Research. 16: 232–242.

Schumaker, N.H.; Ernst, T.; White, D.; Baker, J.; Haggerty, P. 2004. Projecting wildlife responses to alternative future landscapes in Oregon's Willamette Basin. Ecological Applications. 14: 381–400.

Shafer, S.L.; Bartlein, P.J.; Thompson, R.S. 2001. Potential changes in the distributions of western North America tree and shrub taxa under future climate scenarios. Ecosystems. 4: 200–215.

Sharma, S.; Couturier, S.; Côté, S.D. 2009. Impacts of climate change on the seasonal distribution of migratory caribou. Global Change Biology. 15: 2549–2562.

Sohngen, B.; Mendelsohn, R.; Sedjo, R. 2001. A global model of climate change impacts on timber markets. Journal of Agricultural and Resource Economics. 26: 326–343.

Spehn, E.M.; Hector, A.; Joshi, J.; Scherer-Lorenzen, M.; Schmid, B.; Bazeley-White, E.; Beierkuhnlein, C.; Caldera, M.C.; Diemer, M.; Dimitrakopoulos, P.G.; Finn, J.A.; Freitas, H.; Giller, P.S.; Good, J.; Harris, R.; Högberg, P.; Huss-Danell, K.; Jumpponen, A.; Koricheva, J.; Leadley, P.W.; Loreau, M.; Minns, A.; Mulder, C.P.H.; O'Donovan, G.; Otway, S.J.; Palmborg, C.; Pereira, J.S.; Pfisterer, A.B.; Prinz, A.; Read, D.J.; Schulze, E.D.; Siamantziouras, A.-S.D.; Terry, A.C.; Troumbis, A.Y.; Woodward, F.I.; Yachi, S.; Lawton, J.H. 2005. Ecosystem effects of biodiversity manipulations in European grasslands. Ecological Monographs. 75: 37–63.

Stabach, J.A.; Laporte, N.; Olupot, W. 2009. Modeling habitat suitability for grey crowned-cranes (*Balearica regulorum gibbericeps*) throughout Uganda. International Journal of Biodiversity and Conservation. 1: 177–186.

Stein, S.M.; McRoberts, R.E.; Nelson, M.D.; Mahal, L.M.; Flather, C.H.; Alig, R.J.; Comas, S. 2010. Private forest habitat for at-risk species: Where is it and where might it be changing? Journal of Forestry. 108(2): 61–70.

Sulzman, E.W.; Poiani, K.A.; Kittel, T.G. 1995. Modeling human-induced climate change: a summary for environmental managers. Environmental Management. 19: 197–224.

Tewksbury, J.J.; Huey, R.B.; Deutsch, C.A. 2008. Putting the heat on tropical animals. Science. 320: 1296–1297.

Thomas, C.D.; Lennon, J.L. 1999. Birds extend their ranges northwards. Nature. 399: 213.

Thompson, M.; Adams, D.; Johnson, K.N. 2009. The albedo effect and forest carbon offset design. Journal of Forestry. 107: 425–431.

Thuiller, W. 2007. Climate change and the ecologist. Nature. 448: 550–552.

Tilman, D.; Reich, P.B.; Knops, J.; Wedin, D.; Mielke, T.; Lehman, C. 2001. Diversity and productivity in a long-term grassland experiment. Science. 294: 843–845.

Tryjanowski, P.; Kuzniak, S.; Sparks, T. 2002. Earlier arrival of some farmland migrants in western Poland. Ibis. 144: 62–68.

U.S. Department of Agriculture [USDA]. 2006. Conservation reserve program longleaf pine initiative. http://www.fsa.usda.gov. (10 May 2010).

U.S. Department of Agriculture, Natural Resources Conservation Service [USDA NRCS]. 2010. Wetlands reserve program. http://www.nrcs.usda.gov/programs/wrp/. (12 May 2010).

Valiela, I.; Bowen, J.L. 2003. Shifts in winter distribution in birds: effects of global warming and local habitat change. Ambio. 32: 476–480.

van Asch, M.; Visser, M.E. 2007. Phenology of forest caterpillars and their host trees: the importance of synchrony. Annual Review of Entomology. 52: 37–55.

Volney, W.J.A.; Fleming, R.A. 2000. Climate change and impacts of boreal forest insects. Agriculture, Ecosystems and Environment. 82: 283–294.

Wake, D.B. 2007. Climate change implicated in amphibian and lizard declines. Proceedings of the National Academy of Sciences. 104: 8201–8202.

Walter, W.D.; Vercauteren, K.C.; Gilsdorf, J.M.; Hygnstrom, S.E. 2009. Crop, native vegetation, and biofuels: response of white-tailed deer to changing management priorities. Journal of Wildlife Management. 73: 339–344.

Warren, M.S.; Hill, J.K.; Thomas, J.A.; Asher, J.; Fox, R.; Huntley, B.; Roy, D.B.; Telfer, M.G.; Jeffcoate, S.; Harding, P.; Jeffcoate, G.; Willis, S.G.; Greatorex-Davies, J.N.; Moss, D.; Thomas, C.D. 2001. Rapid response of British butterflies to opposing forces of climate change and habitat change. Nature. 414: 65–69.

White, E. 2010. Woody biomass for bioenergy and biofuels in the United States—a briefing paper. Gen. Tech. Rep. PNW-GTR-825. Portland, OR: U.S. Department of Agriculture, Forest Service, Pacific Northwest Research Station. 45 p.

White, E.; Alig, R.J.; Haight, R.G. 2010. The forest sector in a climate changed environment. In: Alig, R.J.; tech. coord. Economic modeling of effects of climate change on the forest sector and mitigation options: a compendium of briefing papers. Gen. Tech. Rep. PNW-GTR-833. Portland, OR: U.S. Department of Agriculture, Forest Service, Pacific Northwest Research Station: 1–35. Chapter 1.

Williams, D.W.; Liebhold, A.M. 2002. Climate change and the outbreak ranges of two North American bark beetles. Agricultural and Forest Entomology. 4: 87–99.

Wilson, W.H., Jr. 2007. Spring arrival dates of migratory breeding birds in Maine: sensitivity to climate change. The Wilson Journal of Ornithology. 119: 665–677.

Wilson, R.J.; Guitiérrez, D.; Guitiérrez, J.; Martínez, D; Agudo, R.; Monserrat, V.J. 2005. Changes to the elevational limits and extent of species ranges associated with climate change. Ecology Letters. 8: 1138–1146.

Chapter 2: Land Value Changes and Carbon Sequestration as an Ecosystem Service in a Climate-Changed Environment

Ralph J. Alig, Edward A. Stone, and Eric M. White

Introduction

Global climate change induced by anthropogenic release of greenhouse gases (GHGs), including carbon dioxide (CO_2), may modify the growth and geographic distribution of forests, as well as productivity of competing land uses. Impacts on agriculture and forests could arise from increases in atmospheric CO_2 concentrations, change in temperature regimes, and variations in patterns of rainfall over the course of a year. Shifts in global climatic conditions could affect agricultural crop and forest growing conditions, and hence land values for those uses. Climate change policies may in turn affect land and resource markets, thereby modifying land values, land use, and forest cover distribution. A broad examination of such changes necessitates assessing multiple markets, including those for carbon sequestration as an ecosystem service, especially if policies arise to promote carbon sequestration as a mitigation activity.

Carbon credits and carbon markets are often key components of proposed national and international attempts to mitigate the growth in concentrations of GHGs. At the same time as policy is causing land-use changes, climate change itself is expected to cause changes to environmental conditions, influencing the characteristics and growth of forests in rural and urban settings, as well as agricultural production and that of other competing land uses. The focus of this paper is how climate change may change land values at broad scales, both from the direct impact from future altered climate conditions and from policies formulated to address climate change and increase the supply of tradable carbon credits. We summarize findings from U.S. studies, to illustrate examples of land value changes in response to climate change and associated policies. Understanding land use changes and land values in the face of climate change is important for creating efficient strategies for the mitigation of climate change at local, regional, and global levels. We focus on the domestic situation, and readers interested in land use and climate change connections for other countries are referred to Rai (2009) and works for other countries.

Changes in Land Values

Land resources such as forest land can produce a future flow of recurring net returns or land rents. Land values typically represent the present value of such net returns, dependent on estimating the future net returns and the choice of an appropriate capitalization rate (Barlowe 1978). Theoretical models of land-use change are grounded in the landowner's dynamic optimization problem. A landowner is assumed to choose the land use that maximizes the net present value (NPV) of the stream of rents to his or her land. A basic model posits that a change in land use occurs whenever the NPV of the rents to an alternative land use exceeds the NPV of the rents to the current land use by more than the cost of conversion.[1] Over time, changes in supply and demand for land affect relative land rents (Alig and Plantinga 2004). The rents accruing to any particular land use are determined by a host of factors such as land productivity, location, and demographics (e.g., population or income changes). Usefully, land rents reflect the interaction of these determinants of land values, which are frequently difficult to observe. Many empirical analyses of land-use change use the rents accruing to different land uses to model and forecast land-use transitions. Given the infinite time horizon, our lack of perfect knowledge and foresight complicates the determination of future rents. To the extent that we are interested in how the allocation of land uses may change with possible carbon revenues, we are also interested in how land values as a key measure may be affected by carbon markets.

Some empirical studies also incorporate market values of land, which when expressed as a stream of rents are conceptually similar to net rents. However, market values may incorporate expectations about the future differently than do calculated net returns. First, market values may incorporate expectations of changes in land use. For example, forest-land prices anticipate future development close to urbanizing areas (e.g., Wear and Newman 2004). The market value of land can be thought of as having two components: (1) the value of present assets (e.g., structures or standing crops or timber) and (2) the speculative value of future land uses and assets that incorporates expectations about a wide range of variables including commodity prices, development patterns, government policies such as for carbon sequestration, and regional population and income levels. Inconsistencies between real estate values and rents for the current land use can be indicative of an expected land-use change. Second, the time horizon for which expectations are formed is likely to be longer for market values.

Projected shifts in land use reflect economic and soil quality conditions in the case of forest and agriculture, whereas conversion of forest land to development is influenced significantly in many cases by location. Increases in population and the demand for land can give rise to incentives to bring less fertile lands into use. For

---

[1] In the presence of risk or uncertainty the landowner's decision problem is more complicated, particularly where the cost of conversion is large or the land-use change is irreversible.

example, with expanded development, growing scarcity of forest land acts to raise timber prices. If timber prices and associated timber-related incomes rise enough, it may be profitable for some owners to intensify timber management on some lands and also afforest additional land. At the same time, changes in transportation costs may affect the extent of the area within which forest products can be profitably produced. Such changes affect land values in forestry.

Urban and developed uses typically sit on top of the economic hierarchy of land uses, with rents often at least an order of magnitude higher than those for forest land (Alig and Plantinga 2004). Alig and Plantinga (2004) reported that the ratio of average value of land in urban use compared to forest use is approximately 87 in the Southeast and 111 in the Pacific Northwest Westside region. The highest forest values on a county basis in the Pacific Northwest Westside are about 0.04 the lowest urban values and 0.007 the highest urban value. The ratios in the Southeast are roughly similar; however, overall, the Pacific Northwest Westside has much larger land values in forest use and urban use. The Pacific Northwest Westside has 14 counties with land values at least $200,000 per acre ($494,200 per ha) in urban use and $2,000 per acre ($4,942 per ha) in forest use, and the Southeast has none. Such higher rents mean that determinants of land-use transitions in many cases are demand-side factors pertaining to developed uses, such as population and income. Thus, in interface areas such as metro or urbanizing locations, the economic hierarchy of land uses suggests that, in land markets, development-related land-use factors tend to strongly dominate forestry-related ones.

Ecosystem Service Markets

In this section we discuss the implications of climate change for ecosystem service provision and examine how changes in ecosystem service provision affect land values.

Defining ecosystem services—

Broadly defined, ecosystem services are the benefits people derive from the natural processes that sustain ecosystems. They can be generally catalogued into four broad areas: supporting, regulating, provisioning, and cultural (MEA 2005). Supporting services include basic ecosystem functions like soil formation, and provisioning services are important sources of food and fiber. Regulating services help control climate change through carbon sequestration. Cultural services include recreation and education. So ecosystem services comprise both goods and processes.

Although the concept of ecosystem services is inherently based on the value or importance to humans, the underlying ecological structure and processes involve complexities that complicate quantifying these services at different scales. In some cases, the linkage between structure, processes, and resulting services is fairly straightforward. For example, the degree to which a specific plant community can support a given wildlife population can be determined directly by measuring community attributes such as species composition, height, and age. Other services, such as improving water quality by converting nitrate to nitrogen gas through denitrification, are controlled by more complex interactions among multiple ecosystem attributes that are more difficult to measure (for example, carbon, redox status, denitrifier population, and temperature). These relationships are also altered by temporal and condition gradients, which result in dynamic processes and significant variability across and within different ecosystems, making them difficult to measure and quantify at large spatial scales. Ecological production function models based on biophysical inputs are often used to produce spatial estimates of specific services (Nelson et al. 2009).

An ecosystem service differs from an ecosystem function because a service takes the human context into account. For example, consider a wetland that provides, among other benefits, flood control. The level of this ecosystem function is the wetland's capacity to absorb rising flood waters and gradually release them over time. The level of ecosystem service is the benefit of that flood control; thus it depends on what lies downstream. A wetland that contributes to flood control for a town or city provides a higher level of service than a wetland protecting undeveloped land. Location is important.

Ralph Alig

Efforts to reduce the divergence between the private provision of ecosystem services and socially optimal amounts has involved payments for ecosystem services.

Theoretically, ecosystem services tend to be under supplied by the private sector because they often represent a positive externality. Although a landowner could manage to increase the provision of an ecosystem service, little economic incentive exists to do so because the benefits of these management practices do not accrue solely to the landowner and little increased revenue may result from increased provision of that service. Ultimately, the provision of ecosystem services flowing from private lands is less than what is desirable to society collectively. Payments for ecosystem services (PES) constitute one strategy to increase the provision of ecosystem services via an economic incentive. Examples include government programs (e.g., the U.S. Conservation Reserve Program [CRP]), eco-labeling and certification programs (if the rents flow to farmers), and purchase of conservation easements (e.g., by The Nature Conservancy for increasingly prevalent land trusts). The provision of ecosystem services from private lands could also be increased through regulations (e.g., best management practices for forestry). In this context, regulations mandate desirable management practices, and PES programs incentivize desirable management practices.

The goal of PES programs is to increase, on the margin, the provision of ecosystem services by paying landowners to adopt beneficial management practices. Carbon sequestration is a prominent example in this context. Sequestering carbon should mitigate negative climate change impacts for society at large. In the absence of some sort of carbon market, landowners will sequester too little carbon because they do not reap the full benefits of sequestration. A carbon market or allowance system incentivizes private land sequestration by making sequestration pay. Given the opportunity to increase revenues, many landowners may change behavior to increase carbon sequestration to levels that are more socially desirable.

The level of ecosystem service provision from a parcel or a region depends on site characteristics (e.g., soil, typical climate conditions—average rainfall, typical growing season, etc.), stochastic shocks (e.g., variable weather patterns, natural disasters), and land management practices. Altering management practices should alter the provision of services, holding other factors constant. Thus, the relationship between ecosystem service and ecosystem function varies with management practices.

Effects of changing climate—
In general, changing climate conditions can be expected to result in changes to site characteristics and stochastic shocks—changes to ecosystem function. Thus, for a given site, the average conditions will change as will the frequency and severity of departures from that average. The implications for ecosystem service provision depend on management response. With static management, ecosystem service and

function should move together. However, human adaptation to changing conditions can mitigate the ecosystem service impacts. For example, Sohngen and Mendelsohn (1998) showed that climate change impacts on forestry in North America would be significantly altered by management practices including salvaging dead trees and speeding up the transition between species. In fact, their results suggest gains for the U.S. forestry sector from climate change. Thus, climate change interacts with human behavior, determining the outcome in terms of ecosystem services in the process. Although there is a general scientific consensus that climate change will be costly overall, the impacts on particular sectors or services may be positive in some regions.

Changes in land values from ecosystem services provision—
But how does ecosystem service provision affect land value, and thus how will climate-induced changes in ecosystem service provision affect land value? First, climate change affects land value through changes in productivity. In forestry and agriculture, site productivity drives land values and rental rates. So as climate change affects productivity, land value changes. Over time, land tends to move to the highest value use. Thus, if we see land moving from forestry to agriculture, we can surmise that returns to agriculture currently outstrip returns to forestry.

However, returns to agriculture and forestry do not depend solely on productivity of traditional products. Returns also depend on payments landowners receive from other sources (e.g., government conservation programs, conservation groups, carbon markets). Payments for ecosystem service schemes enhance the returns to a particular land use or management strategy, rendering that strategy relatively more attractive. Recent years have seen significant expansions of PES schemes. Government expenditures have increased steadily in both the United States and the European Union (USDA ERS 2006), and the involvement of conservation groups and institutional actors (such as the Chicago Climate Exchange) is also on the rise. Payment for ecosystem services programs essentially create markets for services that previously had no markets. As more and more markets come online, this should push land values up. Indeed, empirical evidence shows that the CRP, a PES program that pays farmers to remove environmentally sensitive land from production, has resulted in higher values for both agricultural and developed land (Wu and Lin 2010). Simultaneously, the share of land value attributable to ecosystem services should rise, assuming all other things being held constant.

Growing concern over climate change (and other environmental problems) is a major driver of PES programs, which could affect land value. Carbon sequestration is a particularly important ecosystem service in this context. With a carbon-trading scheme in place, landowners will have the option of managing to enhance carbon sequestration and selling this enhanced sequestration potential in the form of carbon credits. Agriculture and forestry differ in terms of carbon sequestration potential, so the introduction of a carbon market affects relative returns, land value, and land use.

Of course, carbon sequestration rates, similar to other ecosystem services, will vary with climate change. The direction of this movement is important. If carbon sequestration potential rises with increased CO_2 concentration, we have a positive feedback loop. Higher CO_2 concentrations could lead to enhanced sequestration rates. Holding land use constant, the effect of sequestration on climate is self-limiting, with a limit on how long the positive feedback would be significant or exist. If, on the other hand, sequestration potential falls with increased CO_2 concentration, we have a negative feedback loop. Higher concentrations lead to lower sequestration, so concentrations continue to rise.

In either case, the market response will enhance the feedback effects. If carbon sequestration potential rises with CO_2 concentration, the return from using a given parcel to sequester carbon also rises. On the margin, some land moves into carbon sequestration, dampening climate change further. If sequestration falls as concentration increases, the return from using a parcel to sequester carbon also falls. Thus, on the margin, some land will move to other uses, contributing to further increases in CO_2 concentration.

Given the economic hierarchy of land uses, the addition of carbon-related revenues associated with that ecosystem service are not likely to cause major changes in land allocation between developed uses and forestry in interface areas or urbanizing areas. However, adding carbon-related revenues to the future stream of net returns for forestry may boost benefits enough to change the allocation of land between forestry and agriculture in some regions. Agricultural land is afforested in the study by Alig et al. (2010a) because expected incomes per hectare under forest use rise above the expected returns per hectare associated with continuing agricultural use. Increasing financial incentives to plant trees, where plantations store multiple metric tons of forest carbon per hectare, would boost forest incomes in some cases well beyond land rental rates from agriculture. As an example, afforested cropland in mixed hardwood forest in the Corn Belt region is estimated to sequester approximately 60 metric tons of forest ecosystem carbon per ha by age 20 (Birdsey 1992). A $15 price per metric ton of CO_2 is equivalent to about $55 per metric ton of carbon. At this price, the carbon payments from the hardwood stand would amount to $8,154 per hectare ($3,300 per acre) as a lump sum at age 20. As an equal annual payment over the same 20-year period, that would be equivalent to about $274 per hectare ($111 per acre) per year (at a 4 percent real discount rate).

The $15 price for CO_2 could result in several hundred thousand more hectares of afforestation in the Corn Belt region as compared to business as usual (Alig et al. 2010a). If that price per metric ton is doubled to $30, then about 2.5 million more hectares could be afforested with carbon markets.

Other ecosystem services also provide benefits, and some values could be estimated using techniques such as the benefit transfer method (e.g., Johnston and Rosenberger 2010). However, unless the amounts are very large, they are not likely to tip the land allocation balance between forest and developed uses. Further, unless there is some sort of market or PES scheme, enhanced ecosystem services may enhance landowner utility, but may not be as likely to be capitalized in land value. One example of another service is nitrogen uptake, sometimes discussed alongside carbon sequestration. In addition to benefits, analysts need estimates of costs for providing ecosystem services, such as converting agricultural land to wetlands. Considering both benefits and costs allows for net benefits to be estimated, including examining the present values of benefits and costs. Although costs of a mitigation activity may be significant, values of the marketable commodities along with the potential ecosystem services values could be even greater. Thus, depending on the assumptions of the benefits to be included, the return on investment in the mitigation activity could be significant.

Afforestation on erodible or other environmentally sensitive agricultural land can enhance ecosystem services and co-benefits, such as enhanced wildlife habitat, in addition to contributing to climate change mitigation.

Explicitly recognizing the complex relationships between ecosystem structure, processes, and services is critical to understanding the potential ancillary effects of carbon sequestration strategies. Any change, either anthropogenic or naturally occurring, that affects structural components such as plant community composition or processes such as nutrient cycling, will affect the quality, quantity, and types of services produced from that ecosystem. Quantification of the effect is a difficult task, as some services, such as biodiversity, can be both a cause of ecosystem functioning and a response to changing management activity. Thus, effects of carbon-specific components are hard to separate. Another problem is that the responses of multiple services to specific carbon-related management activities have not been well studied. Nelson et al. (2008) concluded that policies aimed at increasing carbon sequestration did not necessarily increase species conservation and that highly targeted policies were not necessarily better than more general policies. The study by Nelson et al. (2008) demonstrates the likelihood that many of the possible management activities and sequestration strategies will affect ecosystem services of direct importance to landowners and managers. For example, afforestation designed to increase carbon sequestration will alter migratory bird habitat depending on the location and species composition of the forest, so ecological tradeoffs will be involved.

Understanding prospects for carbon markets—

The promise of ecosystem service markets greatly depends on the particular circumstances of program implementation, including what services are to be traded and whether they are amenable to trading, the ability to enact and enforce regulation sufficient to induce trading, and how expected program results compare with those likely to arise from other conservation policy approaches (Kline et al. 2009). In the case of tradable forest carbon credits, still needed from an institutional viewpoint is a cap on the amount of carbon emissions. In an unregulated world, carbon emissions exemplify the problem with externalities and public goods. Carbon emitters can carry out production without fully internalizing the social costs of the carbon they emit. All producers contribute to total atmospheric carbon and the resulting negative externality of global climate change. The lack of property rights to the atmosphere provides producers little incentive not to pollute, because they need not pay for the right to emit carbon.

With a regulatory approach, one proposal is to reduce total emissions by establishing a uniform cap on the emissions that individual producers can emit. Thus, the total cap is the uniform cap times the number of producers. A uniform cap is not the least-cost option for achieving the total cap because different producers face different marginal pollution control costs. The least-cost or cost-effective option equalizes marginal control cost across producers. The intent of creating a carbon market is to provide a compliance option for heavy emitters—the ability to purchase allowances to maintain production without having to reduce their carbon emissions. Increasing the flexibility of compliance options for emitters in this way equalizes marginal cost and lowers the cost of achieving the total cap. Over time, the regulatory agency could ratchet down the overall cap on emissions to gradually reduce total carbon emissions.

In this carbon market, forest owners could conceivably participate by selling offsets in the form of tradable carbon credits, increasing the supply of emission or damage allowances in a market. The forest carbon owners are performing activities that offset damage caused by others (Kline et al. 2009). Landowners might be given the right to sell emissions allowances based on forest management or afforestation activities that increase forest carbon storage on their lands above some predetermined baseline.

Next, we summarize aspects of land-use changes, land markets, and land values based on modeling studies. The opportunity for exurban and urban developments to influence the supply of terrestrial carbon sequestration is outside the scope of this chapter, and interested readers are referred to Alig et al. (2004) and White et al. (2009) for examples of development area projections. In this chapter, we focus on land-use change within a climate change context, although land-use change can have many implications for a landscape and society (Alig et al. 2004). Land-use changes can also affect product prices (White et al. 2010), wildlife habitat (see chapter 1), and many other parts of ecosystems and the interconnected economic system.

Examples of Land Value Projections Using the FASOM-GHG Modeling System

Global questions involving forestry's and agriculture's potential contributions to climate change mitigation are framed within a context of increased demands for cropland, forage, and wood products to help feed and house an additional 3 billion people globally by 2050, increased land demand for bioenergy production, and millions of hectares of land needed to house another 125 million U.S. residents by midcentury (Alig et al. 2010b). Here we illustrate how the Forest and Agricultural Sector Optimization Model—Greenhouse Gas (FASOM-GHG) modeling system is used to examine such linked forestry and agricultural issues. Examples of components in FASOM-GHG's integrated mitigation analysis include (a) simultaneous consideration of the agricultural (including both crops and livestock) and forestry sectors; (b) discount rate; (c) modeling of impacts on traditional industries; (d) dynamic modeling of forest stand growth and investment; (e) GHG accounting, including non-CO_2 gases; (f) GHG (e.g., CO_2) prices; (g) modeling conversion of forest and agricultural lands to developed uses (e.g., urban); (h) examination of bioelectricity as part of bioenergy analyses; and (i) land-use changes between the forestry and agricultural sectors, among other components.

Examining land use change and climate change can involve looking at effects of climate change on forestry and agricultural activities, adaptation activities, mitigation activities, and the interplay among all three. Within the narrower mitigation modeling, such modeling should account both for the GHG impacts of substituting bioenergy for fossil fuels and any GHG changes owing to land-use change—so we should get the net GHG impact of any change to bioenergy production. Increasingly, modeling could point out exactly the kinds of perverse GHG incentive results described in some of the literature. For example, some types of subsidies could lead to reduced carbon sequestration as an ecosystem service. Including both the agricultural and forest sectors, modelers can directly recognize the tradeoffs between the needs to reduce GHG emissions and expand food production. Models such as the Forest and Agricultural Optimization Model (e.g., Alig et al. 2010a) demonstrate the forest environmental impacts of different types of policies. Tradeoffs with food and environment will occur, and we can use models such as FASOM-GHG to investigate such tradeoffs along the path to reducing net GHGs.

In the next section, we examine studies that project how future climate conditions may impact land values, as well as how policies proposed in response to climate change may affect forest-land values. This involves using models that link expected forest changes from climate change to the inputs and parameters used in forest sector economic models (White et al. 2010). A linked model of the forest and agricultural sectors simulates land-use competition between the two sectors, with resultant impacts on land allocation and land values in the sectors.

Modeling Changes From Climate Change

Most studies thus far have assumed that the total area of land in agriculture and the total area of land in forests will remain about the same even as climate changes. For example, Sohngen and Mendelsohn (1998) valued the impact of climate change as a large-scale ecological change in the U.S. timber market, without a mechanism for accounting for any productivity-related shifts between forestry and agriculture. Although the total area of land in agriculture and forestry may indeed remain relatively constant over time, climate change could alter the distribution of land uses over time. Given that forest areas could be affected and the potential human and social consequences of these impacts, it is important to consider and assess the implications of climate change on the distribution of U.S. land uses.

Differential impacts of climate change in agriculture and forestry could lead to land-use shifts as one possible adaptation strategy by landowners. For example, if climate change results in relatively higher agricultural productivity per hectare, some hectares may be converted from forests to agricultural use. Such changes would alter the supply of products to national and international markets, changing the prices of forest products and the economic well-being of both producers and consumers. Conversely, if climate change renders forestry more attractive than agriculture in terms of yields and production costs, land could shift from agriculture to forestry as these two sectors adjust. Given that the agricultural and forestry sectors sometimes compete for the same land, shifts in productivity of agricultural land could affect the ultimate distribution of forest land, and vice versa.

Using four climate change scenarios from the national assessment of climate change, Alig et al. (2002) found that less forested area was projected under four climate scenarios relative to the base case (no climate change). Furthermore, less cropland and more pasture land were projected to convert to forests under all scenarios. In general, climate change was projected to increase the overall supply of timber, causing timber prices to fall and reducing forest-land values. The net conversion of forest land to agricultural use was prompted by aggregate increases in forest productivity relative to agriculture, with price signals in markets leading to adjustments. Although climate change is likely to affect the margin between forestry and agriculture differently in specific locations, aggregate productivity changes in forestry appear larger overall than aggregate productivity changes in agriculture.

Ralph Alig

Land values represent key links between forestry and agriculture land markets.

Other studies have examined impacts of climate change on agricultural yields. For example, Schlenker and Roberts (2008) estimated impacts of climate change on crop yields, considering the importance of nonlinear temperature effects. However, they did not consider any land use shifts involving forestry that might arise from impacts on productivity in the two sectors. Similarly, Sohngen and Mendelsohn (1998) examined impacts on forest productivity but did not account for land use shift possibilities.

To our knowledge, no study to date has isolated the impacts on forest-land values from carbon sequestration owing to climate change. The earlier version of the FASOM model employed by Alig et al. (2002) did not have carbon-related prices in the objective function. Alig et al. (2002) did note changes in economic welfare by sector associated with the different climate change scenarios, with changes relatively small compared to those associated with policy scenarios.

Economic impacts of climate change on land use distribution could also involve human migration patterns and impact the amount and pattern of urban and developed areas by region. Regional patterns of growth and decline in the United States show shifts in population and property value to more vulnerable areas (van der Vink et al. 1998), and concerns about climate change and severe weather events could alter coastal settlement patterns. A large amount of uncertainty surrounds any such movement of population and land use impacts, and is outside the scope of this paper.

Modeling Changes in Land Values With Policies and Carbon Markets

Forests are cost-competitive in their ability to play some role in U.S. carbon sequestration activities (Richards and Stokes 2004). Cost studies of forest carbon sequestration over more than a dozen years have estimated costs of between about $10 and $150 per metric ton of carbon (Dempsey et al. 2010, Richards and Stokes 2004). Some estimates point to forests being able to sequester between approximately 250 and 500 million metric tons in the United States per year, mirroring the growth of the forest and storage of carbon in the accumulating woody biomass. Richards and Stokes (2004) pointed out that the cost effectiveness of forest-based sequestration depends on the secondary benefits of sequestration (e.g., habitat creation and watershed improvement).

Next, we look at the use of the FASOM-GHG model to investigate changes in values for forestry, agriculture, and tradable carbon credits with carbon-related policies. The FASOM-GHG modeling framework can also be used to examine how expectations of land-base shifts between the agriculture and forest sectors impact potential policy effectiveness. Because market equilibrium is determined with endogenous interaction between the forest and agriculture sectors, varying levels of future land use patterns can be explored by the introduction of constraints.

Alig et al. (2010a) used the FASOM-GHG modeling system to investigate effects of carbon dioxide equivalent (CO_2 e) prices of between $0 and $50 per metric ton. These carbon-related values are added to the objective function in a symmetric fashion. A symmetric system, also called a carbon subsidy-tax system, pays forest owners for increments and "taxes" them for any decrements in carbon stock. Hypothetical policies involving carbon-related prices could cause landowners to undertake changes in forest management as well as afforest additional agricultural land. Although outside the scope of this chapter, climate change policies may involve mitigation actions that promote reduced deforestation. Alig et al. (2010a) summarized findings about such efforts to conserve forest land and the impacts on carbon sequestration and storage and on traditional forest industries.

The CO_2 pricing scenarios reported by Alig et al. (2010a) included a CO_2 price of zero, $25/metric ton, and $50/metric ton. The $25/metric ton price approximates levels estimated by Murray et al. (2005), who estimated that costs of mitigation actions in forestry and agriculture would range from $15 to $25/metric ton of CO_2 mitigated. Their analysis using FASOM model runs indicated that between 10 and 25 percent of current U.S. GHG emissions could be offset through a combination of actions in forestry and agriculture, including reduced cropland tillage, afforestation, improved forest management, improved nutrient management, manure management, and bioenergy production (Murray et al. 2005). They also simulated a scenario with $50/metric ton to investigate effects of a higher CO_2 price.

Alig et al. (2010a) reported that soil expectation values (SEV), as one proxy for effect on land values, on average almost doubled with a $25 CO_2 price compared to the base case with no carbon price. These are "bare" forest-land values. Moving to a $50 CO_2 price increased the SEV values by more than 2.5 times compared to the base case.

Although these are substantial increases, the increases in land value are a magnitude of order smaller than the differential between forest land and developed use values reported by Alig and Plantinga (2004). For example, in the Southeast, SEV timberland values in the Alig et al. (2010a) base run were approximately $500 to $750 per hectare (approximately $200 to $300 per acre) on average. Doubling or tripling such land values would leave them short of $2,500 per hectare ($1,000 per acre), whereas Alig and Plantinga (2004) reported average urban values in the Southeast of more than $90,000 per hectare ($36,000 per acre), an average of more than 30 to 40 times as much.

Therefore, although carbon-related revenues are not likely to appreciably affect the land competition between forestry and developed uses in urbanizing areas, such carbon revenues can conceptually cause switches in land use between forestry and agriculture. As shown in figure 2-1, with afforestation area sensitive to CO_2 prices, projected net change in forest area involving land exchanges with agriculture (afforestation area minus area deforested to agriculture) summed over the first 50 years of the projection was positive with a \$20 CO_2 price, in the Alig et al. (2010a) study. In contrast, net change levels without a CO_2 price are negative. Afforestation dominates deforestation to agriculture early in the projection with \$25 CO_2, but then is exceeded by such deforestation for most of the projection period. The degree of land use change associated with a particular carbon price will vary by region. For example, Lewandrowski et al. (2004) showed that the Pacific Northwest region is relatively less responsive to incentive payment levels for afforestation than the Southeast/Appalachia region.

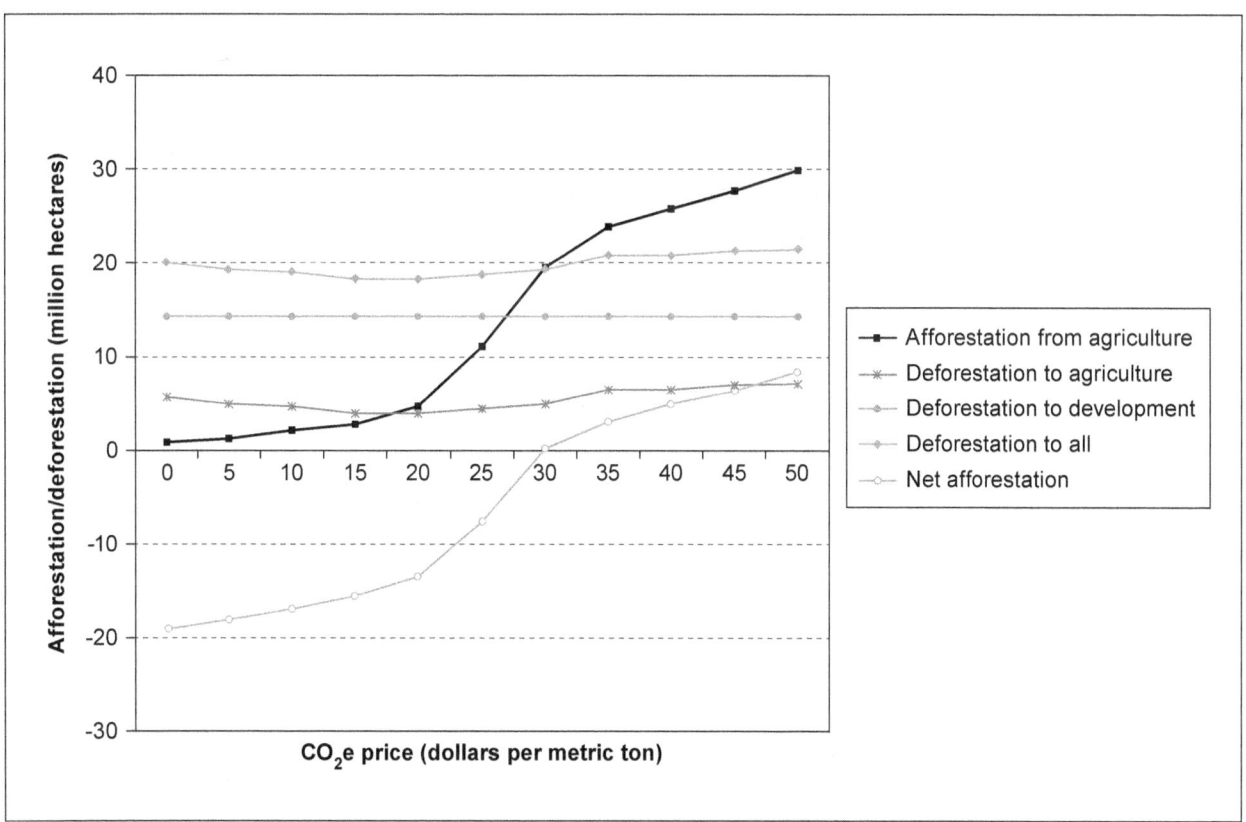

Figure 2-1—Net projected afforestation under different carbon dioxide prices. Adapted from Alig et al. 2010a.

Projected area of afforestation follows a generally declining trend over time in the Alig et al. (2010a) national study. Afforestation area with \$50/metric ton CO_2 averages 0.92 million hectares for the first 25 years of the projection, and then drops to 0.31 million hectares over the next 25 years. The highest CO_2 prices boost forest area by 55 percent compared to the base case, especially in the South and North. Doubling the CO_2 price from \$25 to \$50 per metric ton boosts forest area in 2050 by 17 percent.

Deforestation for conversion to agricultural use is sensitive to the CO_2 price assumption. Deforestation to agriculture is reduced to about one-fifth the base area amount with \$25 CO_2 prices (Alig et al. 2010a). With \$50 CO_2, the amount of deforestation to agriculture drops essentially to zero over the first two projection decades. However, even with \$50 CO_2 prices, the deforestation for agricultural use does increase later in the projection period (Alig et al. 2010a), given the expanded forest area and changes in land prices across the two sectors.

Altering the assumption of no loss of forest to development has a smaller effect than having CO_2 payments for landowners (Alig et al. 2010a). Compared to the base level assumption, avoiding deforestation to urban and developed uses can boost forest area in 2050 by about 9 percent, about half the increase in moving from \$25 to \$50 CO_2 prices. The projected base level of forest converted to development is approximately 15 million hectares over 50 years, drawn from exogenous projections for the 2010 Resources Planning Act (RPA) Land Base Assessment by the USDA Forest Service (Alig et al. 2010b; Lubowski et al. 2006, 2008; Plantinga et al. 2007). The largest projected losses are in the South and Northeast, consistent with recent historical trends.

Evidence of Carbon Markets

The section above pertains to hypothetical policies for carbon sequestration. However, what is actually going on with carbon markets in the United States? Currently, there is no binding global agreement to reduce carbon emissions, and no associated carbon market. Climate change is a topic replete with uncertainty and the progress of carbon markets would require a regulatory cap for significant offset values to arise. Ecosystem service markets are one approach among several for ecosystem protection and part of the future course of such markets will depend on recognition of whether greater global ecosystem protection is necessary. The Europeans have had a cap-and-trade market running since 2005. The first few years were experimental. Initially, there was a lot of price volatility and a fundamental imbalance between supply of tradable carbon credits and demand for them. With a more stable and clear regulatory framework, the European carbon trading experiment continues. A mature trading program will provide valuable data and give economists the opportunity to observe interaction between carbon trading policy and a real economy. Future transactions in forest carbon credits could constitute a market of billions of dollars annually and involve potentially large transfers of funds between players in our economy.

In terms of actual transactions, buyers of ecosystem services can already purchase carbon credits on the Chicago Climate Exchange (CCX) to offset activities that emit GHGs. In turn, the CCX buys credits from projects that offset the accumulation of GHGs in the atmosphere. Many forestry projects do just that, and some are already generating income for forest landowners. One example is carbon currently being traded on the CCX. Figure 2-2 shows that carbon prices in recent times have fallen below $1 per metric ton.

A number of organizations are angling to position forest landowners to be able to trade carbon sequestered on their land on the CCX. In October 2007, California Air Resources Board adopted the first standards in the United States for forest-generated, CO_2 emissions reduction projects. AgraGate Climate Credits Corp. is

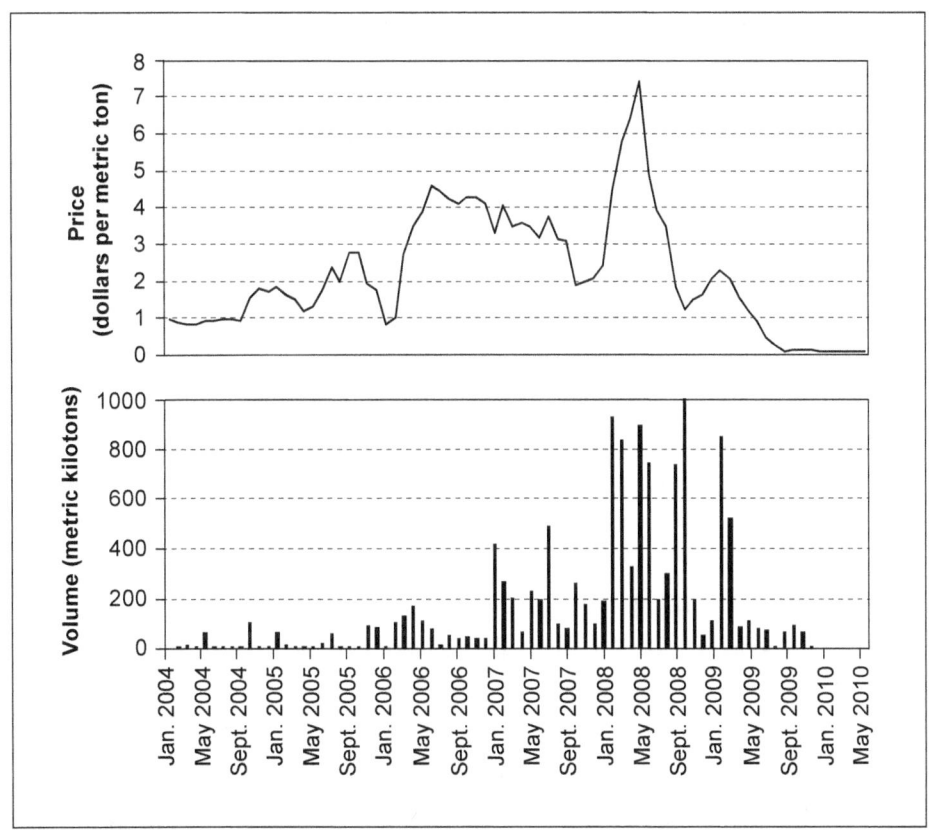

Figure 2-2—Monthly price and volume data for Carbon Financial Instrument (CFI) contracts on the Chicago Climate Exchange (CCX).

now able to provide a Forestry Offset contract on the CCX to forest owners. F&W Forestry Services, Inc. is in a similar position. The Georgia Carbon Sequestration Registry assists in the application of land management practices to sequester carbon and creates new economic opportunities for landowners. The Texas Forest Service received approval as an Authorized Verifier of Forestry Offset Projects for the CCX. FORECON EcoMarket Solutions, LLC announced its successful development, presentation, and approval from the CCX for the first managed forestry carbon offset project for a timber investment management organization. The pilot project covered about 4050 ha (10,000 acres), mostly in Pennsylvania.

However, challenges include the fact that standards lack any sort of national uniformity. There are no compliance-based forest programs in the United States, and monitoring actual implementation of carbon offset programs is challenging. Struck (2010) suggested that some people are buying into projects that are never completed, or paying for ones that would have been done anyway. Some green schemes range from selling protection for existing trees to the promise of planting new ones that never thrive. In some cases, the offsets have consequences that their purchasers never foresaw. Struck (2010) characterized carbon offsets as the environmental equivalent of financial derivatives: complex, unregulated, unchecked, and—in some cases—not worth their price.

Implementation issues also include possible leakage effects, which could result in governments spending large amounts of money while gaining little or no net gain in carbon reductions. Thus, although it is possible to increase carbon sequestration in forests through afforestation or forest management activities, the net effects on overall carbon sequestration may not be as large as anticipated because land markets in areas where forests are unprotected may respond by moving some forests back into agriculture (i.e., deforestation) (Alig et al. 1997). Alig et al. (1997), for example, found an approximate one-to-one correspondence between hectares converted to forests from agriculture and hectares converted the opposite direction, suggesting the need for a comprehensive policy that recognizes leakage. Murray et al. (2004) found similarly large leakage effects for some regions of the United States, and Sohngen and Brown (2004) found smaller, though still potentially substantial, leakage effects for tropical regions. More recent efforts suggest that efficient policies with flux constraints or carbon pricing could provide net sequestration, and that these would increase timber supply both in the short run and long run (Adams et al. 1999, Murray et al. 2005, Sohngen and Mendelsohn 2003).

It is also important for any carbon credit (offset) to prove a concept called additionality. The concept of additionality addresses the question of whether the project would have happened anyway, even in the absence of revenue from carbon credits (see for example, Langpap and Kim 2010). Only carbon credits from projects that are additional to the business-as-usual scenario represent a net environmental benefit. Thus, assessing additionality requires specifying some sort of baseline—what would have happened without carbon markets. Carbon projects that yield strong financial returns even in the absence of revenue from carbon credits, that are compelled by regulations, or that represent common practice in an industry are usually not considered additional, although a full determination of additionality requires specialist review.

Land Value Data Needs

Forest-use valuation is increasingly complicated, as is our understanding and modeling of our environment and economy. The increasing recognition of forest resource values other than timber, such as for tradable carbon credits, is part of a larger societal concern about sustainability of land to provide the goods and services that we as a society demand (Beuter and Alig 2004). Given the fixed amount of land, we need to ask what prices we are willing to pay for those goods and services and how that will affect land values. We need improved databases pertaining to land markets that provide evidence on revealed behavior about what people are willing to actually pay for a bundle of rights necessary to gain access to land that can provide goods and services into perpetuity.

On the empirical side, analyses of land-use change over the past two decades have tended to express land value in terms of the stream of rents, or net returns, accruing to a particular land use. Three general types of data used to estimate net returns are (1) land coverage data, (2) production data, and (3) land value or price data. Land coverage data identify a particular land use at a point in time and may also provide a measure of land quality or suitability for alternative uses (e.g., soil classification). Production data identify the quantities of output associated with a particular land use for a specific geographic area and period. An example of a need for improved production data is afforestation yields on former cropland and pastureland in all regions. Price data identify input and output prices for the production and sale of goods related to a particular land use, such as for forest carbon sequestration. Efforts to improve land valuation for climate change studies would likely benefit most from attention to the latter category. Increased interest in carbon markets has resulted from growing awareness about climate change and the need for mitigation activities, many of which could involve terrestrial carbon sequestration. Ecosystem service markets lately have become a popular topic among environmental policymakers (Kline et al. 2009), and carbon sequestration has received a lot of attention. However, with emerging or nascent markets, many questions remain to be answered. For example, will demand for various ecosystem services be complementary or competitive, and how may pricing of new services affect land use, food prices, and the prospects for biofuels or other bioenergy production?

One information need is for monitoring of emerging forest carbon markets and the relationship to land values. Emerging markets may involve speculative elements and often include nontraditional markets for forest goods and services. Valuation information for these emerging markets may be notably more limited than for commodity-related forest products. Further, forest carbon sequestration opportunities exist on both private and public lands and often are a joint or byproduct of forest management for other objectives.

One example of a database as a candidate to be built upon is Lubowski's (2002) construction of a nationwide database of net return estimates for several major land uses on private lands. It should be noted that the construction of such estimates can be labor intensive and may require a large degree of extrapolation or inference. His database was constructed to support national land-use modeling, and was later used in other studies (e.g., Alig et al. 2010b).

In general, gaps exist in land value data, such as for forest-land values based on actual transactions generally available to the public. For forest-land values, Lubowski (2002) estimated the weighted net present value (NPV) of sawtimber

Ralph Alig

Carbon markets are emerging in some areas, and further expansion could affect forest-based mitigation opportunities.

from different forest types at the county level. The area of each forest type was identified using the USDA Forest Service's Forest Inventory and Analysis data. Timber yields for different forest types were based on existing forest-land estimates developed by Birdsey (1992), and need to be augmented for afforestation possibilities. State-level stumpage prices were obtained from state and federal agencies or from private companies. Where state-level stumpage prices for private forest were unavailable, sales prices from state or national forests were used. Unavailability of nationwide stumpage price data is an impediment to estimating rents. Although Timber Mart-South has historical records of stumpage prices, its coverage is limited to 11 Southern States. In constructing nationwide rents, Lubowski (2002) proceeded state by state to gather reported stumpage prices by surveying various state agencies and private sources. Hence, price data are likely to be inconsistent in terms of methodology and quality (varying both regionally and over time). Furthermore, developing a nationwide database of stumpage prices in this manner is likely to be labor intensive.

Other data sources may be able to complement, approximate, or improve upon more established sources at the regional, if not national, level. Some of the sources identified below are real estate values. Although real estate values will typically be similar to stream of rent values, they cannot be considered perfect substitutes. Real estate values incorporate expectations about future land uses that may differ from the current use. For example, the real estate value of forestry or agricultural land on the fringe of urban development may be well in excess of the NPV of indefinite forestry or agricultural rents because the land value is "ripening" in accordance with buyers' expectations about its impending change to developed land. In this sense, disparities between market land values and rent values may indicate an impending land use change (Kilgore and MacKay 2007, Wear and Newman 2004).

Since 1997, USDA's National Agricultural Statistics Service (NASS) (e.g., 2009) as a public source has published agricultural land value and rental rate data at the state level, but 2009 marks the first time NASS published the information at the county level. For 2009, NASS reports mean county-level cash rents per acre for cropland (irrigated and nonirrigated) and pastureland. The population target for the data is all farms and ranches, and information is collected primarily by telephone interview. The county-level rents are available online in tabular and map formats (see fig. 2-3[2]), providing a readily available proxy for cropland net returns at the county level. State-level cash rents are available from 1994 to present. However, similar data are not available for forest land.

---

[2] The data in figure 2-3, as well as data for nonirrigated rents, are available at http://www.nass.usda.gov/Charts_and_Maps/Land_Values_and_Cash_Rents/index.asp.

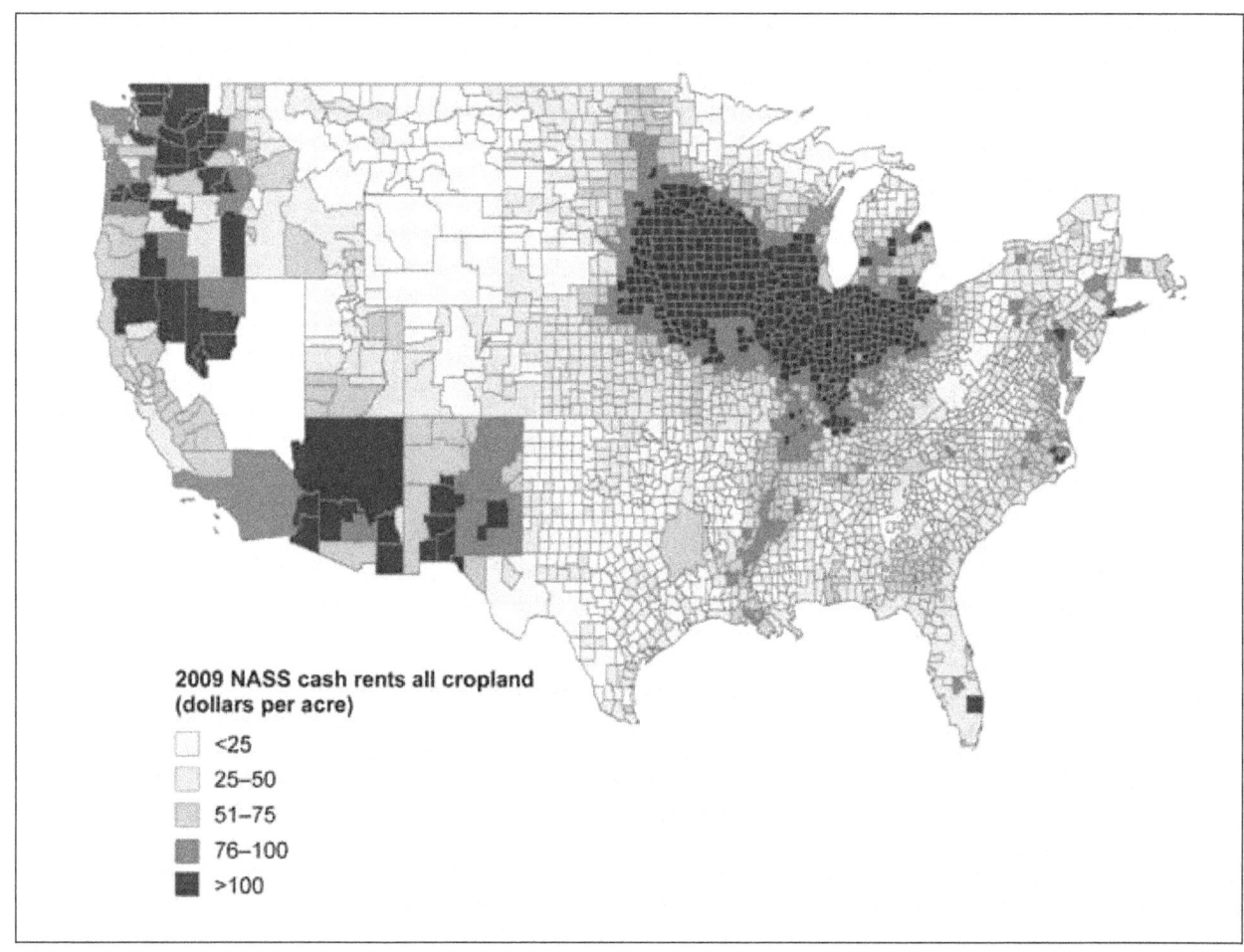

Figure 2-3—2009 National Agricultural Statistical Service (NASS) cash rents, across all cropland. Note that NASS cash rents are based on a survey of farmers. Data are unavailable for parts of Nevada, California, and New England, which appear white in the map.

Conceptually, the cash rent data are the same as the net returns to cropland calculated by Lubowski (2002). To obtain a single rent figure for cropland, irrigated and nonirrigated rents should be weighted proportionally to those lands' acreages. The cash rent data could supplement future updates to the crop rent database by serving as a comparative check or by providing an easy way to expand the database.

In addition to cash rent data, NASS publishes regional and state estimates of average real estate value per acre for irrigated, nonirrigated, and pastured agricultural land. These real estate values could be used to compute annual rents. The disparity between average real estate value and crop rents could be used as a measure of development pressure.

Conclusions and Research Needs

Land prices and other measures of values reflect society's valuation of land for different uses. Modeling changes in land values from climate change and climate change policies can provide valuable information to landowners, managers, and policymakers as society grapples with how to deal with climate change. One facet is how land values may change with increased demand for tradable carbon credits. Transactions in forest carbon credits could constitute a market of billions of dollars annually and involve potentially large transfers of funds between players in our economy.

Forest carbon credits could add significantly to the possible streams of income from timberland. In addition to the growth and harvest of timber, the possibility of selling carbon credits would affect investments in timberland. Forest lands in general sequester (bind up from the atmosphere) substantial amounts of carbon, which in markets could be translated into carbon credits. Revenues from carbon credits would tend to boost forest-land values, especially in comparison to some agricultural alternatives in some regions. One example of a key activity for which better forest and carbon yield data are needed is afforestation of former agricultural lands. In general, we need to better understand how land productivity for different uses may change under future climate change in the various regions. Climate change is being monitored, with some 24 separate indicators (USEPA 2010) showing how climate change affects the health and environment of U.S. citizens. That could be usefully augmented with related indicators about how land productivity is being affected.

Ecosystem service provision can be expected to change with climate. Productivity will change with climate, affecting land value and land use. Additionally, concern over environmental degradation and climate change has increasingly led to the establishment of conservation programs, which can augment land value. A central ecosystem service in the climate change context is carbon sequestration. The direction of the relationship between CO_2 concentration and sequestration determines whether we have a negative or positive feedback loop. In both cases, incorporating the effect on land values tends to reinforce the loop.

If global carbon markets emerge, a comprehensive assessment of the carbon storage and GHG fluxes within the Nation's ecosystems would aid in the evaluation of potential policy and mitigation actions. Such an assessment is inherently complex, as it is geographically broad; covers many different ecosystems, pools, and flux types; and is influenced by many present and future potential conditions (such as climate change, population change, land cover change, ecosystem disturbances, land management activities). If policies adopted to address climate change influence the extent, composition, and management of future forests as well as competing land uses, effects are likely to differ by region. In addition, the nature of the altered climate could vary substantially by region, so that comparative advantages in forest and agricultural production across regions could be affected along with land values.

The policy development process must be supported with integrated, systems analysis. Continued investment in developing and refining the models that provide this type of analysis is critical. The over-arching nature of the climate change issue, which potentially could affect all facets of our lives and work activities (e.g., Fox et al. 2009), means our reliance on integrated models and analysis will continue to increase as policymakers grapple with a confusing array of choices and tradeoffs. Climate change may substantially affect baseline emissions as well as the cost and effectiveness of mitigation options. Although there has been increasing recognition of these linkages, there remains a need for additional model development and analyses to improve our understanding of the implications for mitigation, adaptation, and risk management policy. Demands for bioenergy feedstocks, carbon sequestration, food, fiber, and ecosystem benefits from our land base are increasing. At the same time, the land base faces enormous potential changes in productivity and risks of extreme events. Thus, there are significant needs to improve our understanding of potential effects and to model the complex interactions between competing demands under changing conditions. Risk management has an important role in influencing responses to climate impacts and in developing appropriate mitigation and adaptation policy. There is a need to improve our understanding of how people will change their land use and production practices and the implications for agriculture, forests, and natural ecosystems as they respond to changing risks within the context of policies influencing management choice.

As ecosystem services become scarcer owing to population growth and rising global standards of living (consumption), interest grows in devising markets and other means for procuring these benefits. Research to support carbon policy analyses includes compensation amounts needed as incentives for landowners to participate in tradable carbon credit markets. Land-use change is a key part of global change in that deforestation, urban sprawl, agriculture, and other human influences have substantially altered and fragmented our global landscape. Such disturbances of the land can change the global atmospheric concentration of CO_2, the principal heat-trapping gas, as well as affect local, regional, and global climate by changing the energy balance on Earth's surface. Research questions relate to the relationship between land market activity, land management, and carbon sequestration on the global landscape. For example, given that land management can notably alter future levels of forest carbon sequestration (e.g., Adams et al. 1998), interplay between land markets and markets for products and services can influence the path of carbon sequestration over time. Thus, how might variations in landowner behavior by region, including development of different densities and types, influence the land market outcomes?

Land use change plays a very significant role in climate change mitigation and adaptation, and a land value database could provide both direct and indirect benefits. Land prices or values integrate supply and demand information and could also provide better information on future land-base conditions for climate change analysts. For example, for forest ecosystems, they could provide a thorough and unified description of anticipated change in the valuation by society.

English Equivalents

When you know:	Multiply by:	To get:
Hectares (ha)	2.47	Acres
Metric tons	1.102	Tons

References

Adams, D.M.; Alig, R.J.; McCarl, B.A.; Winnett, S.M.; Callaway, J.M. 1999. Minimum cost strategies for sequestering carbon in forests. Land Economics. 75(3): 360–374.

Alig, R.J.; Adams, D.M.; McCarl, B.A. 2002. Projecting impacts of global climate change on the U.S. forest and agriculture sectors and carbon budgets. Forest Ecology and Management. 169: 3–14.

Alig, R.J.; Adams, D.M.; McCarl, B.A.; Callaway, J.M.; Winnett, S.M. 1997. Assessing effects of mitigation strategies for global climate change with an intertemporal model of the U.S. forest and agriculture sectors. Environmental and Resource Economics. 9: 259–274.

Alig, R.J.; Kline, J.D.; Lichtenstein, M. 2004. Urbanization on the U.S. landscape: looking ahead in the 21st century. Landscape and Urban Planning. 69: 219–234.

Alig, R.J.; Latta, G.S.; Adams, D.M.; McCarl, B.A. 2010a. Mitigating greenhouse gases: the importance of land base interactions between forests, agriculture, and residential development in the face of changes in bioenergy and carbon prices. Forest Policy and Economics. 12(1): 67–75.

Alig, R.J.; Plantinga, A. 2004. Future forestland area: impacts from population growth and other factors that affect land values. Journal of Forestry. 102(8): 19–24.

Alig, R.J.; Plantinga, A.; Haim, D.; Todd, M. 2010b. Area changes in U.S. forests and other major land uses, 1982–2002, with projections to 2062. Gen. Tech. Rep. PNW-GTR-815. Portland, OR: U.S. Department of Agriculture, Forest Service, Pacific Northwest Research Station. 102 p.

Barlowe, R. 1978. Land resource economics. Englewood Cliffs, NJ: Prentice Hall, Inc. 653 p.

Beuter, J.H.; Alig, R.J. 2004. Forestland values. Journal of Forestry. 102(8): 4–8.

Birdsey, R.A. 1992. Carbon storage and accumulation in the United States forest ecosystems. Gen. Tech. Rep. WO-59. Washington, DC: U.S. Department of Agriculture, Forest Service, Washington Office. 51 p.

Dempsey, J.; Plantinga, A.J.; Alig, R.J. 2010. What explains differences in the costs of carbon sequestration in forests? A review of alternative cost estimation methodologies. In: Alig, R.J., tech. coord. Economic modeling of effects of climate change on the forest sector and mitigation options: a compendium of briefing papers. Gen. Tech. Rep. PNW-GTR-833. Portland, OR: U.S. Department of Agriculture, Forest Service, Pacific Northwest Research Station: 87–108. Chapter 4.

Fox, H.; Kareiva, P.; Silliman B.; Hitt, J.; Lytle, D.; Halpern, B.; Hawkes, C.; Lawler, J.; Neel, M.; Olden, J.; Schlaepfer, M.; Smith, K.; Tallis, H. 2009. Why do we fly? Ecologists' sins of emission. Frontiers in Ecology and the Environment. 7: 294–296.

Johnston, R.J.; Rosenberger, R.S. 2010. Methods, trends, and controversies in contemporary benefit transfer. Journal of Economic Surveys. 24(3): 479–510.

Kilgore, M.A.; MacKay, D.G. 2007. Trends in Minnesota's forestland real estate market: potential implications for forestland uses. Northern Journal of Applied Forestry. 24(1): 37–42.

Kline, J.D.; Mazzotta, M.; Patterson, T. 2009. Toward a rational exuberance for ecosystem services markets. Journal of Forestry. 107(4): 204–212.

Langpap, C.; Kim, T. 2010. Literature review: an economic analysis of incentives for carbon sequestration on nonindustrial private forests (NIPFs). In: Alig, R.J., tech. coord. Economic modeling of effects of climate change on the forest sector and mitigation options: a compendium of briefing papers. Gen. Tech. Rep. PNW-GTR-833. Portland, OR: U.S. Department of Agriculture, Forest Service, Pacific Northwest Research Station: 109–142. Chapter 5.

Lewandrowski, J.; Sperow, M.; Peters, M.; Eve, M.; Jones, C.; Paustian, K.; House, R. 2004. Economics of sequestering carbon in the U.S. agricultural sector. ERS Tech. Bull. No. 1909. Washington, DC: U.S. Department of Agriculture, Economic Research Service. 43 p.

Lubowski, R.N. 2002. Determinants of land-use transitions in the United States: econometric analysis of changes among the major land-use categories. Cambridge, MA: Harvard University. 115 p.

Lubowski, R.N.; Plantinga, A.J.; Stavins, R.N. 2006. Land-use change and carbon sinks: econometric estimation of the carbon sequestration supply chain. Journal of Environmental Economics and Management. 51(2): 135–152.

Lubowski, R.N.; Plantinga, A.J.; Stavins, R.N. 2008. What drives land-use change in the United States? A national analysis of landowner decisions. Land Economics. 84(4): 529–550.

Millennium Ecosystem Assessment [MEA]. 2005. Ecosystems and human well-being: current state and trends. Washington, DC: Island Press. 143 p.

Murray B.C.; McCarl, B.A.; Lee, H. 2004. Estimating leakage from forest carbon sequestration programs. Land Economics. 80(1): 109–124.

Murray, B.C.; Sohngen, B.L.; Sommer, A.J.; Depro, B.M.; Jones, K.M.; McCarl, B.A.; Gillig, D.; DeAngelo, B.; Andrasko, K. 2005. Greenhouse gas mitigation potential in U.S. forestry and agriculture. EPA-R-05-006. Washington, DC: U.S. Environmental Protection Agency, Office of Atmospheric Programs. 134 p.

Nelson, E.; Mendoza, G.; Regetz, J.; Polasky, S.; Tallis, H.; Cameron, D.R.; Chan, K.M.A.; Daily, G.C.; Goldstein, J.; Kareiva, P.M.; Lonsdorf, E.; Naidoo, R.; Rickett, T.H.; Shaw, M.R. 2009. Modeling multiple ecosystem services, biodiversity conservation, commodity production, and tradeoffs at landscape scales. Frontiers in Ecology and the Environment. 7: 4–11.

Nelson, E.; Polasky, S.; Lewis, D.J.; Plantinga, A.J.; Lonsdorf, E.; White, D.; Bael, D.; Lawler, J.J. 2008. Efficiency of incentives to jointly increase carbon sequestration and species conservation on a landscape. Proceedings of the National Academy of Sciences. 105: 9471–9476.

Plantinga, A.; Alig, R.J.; Eichman, H.; Lewis, D.J. 2007. Linking land-use projections and forest fragmentation analysis. Res. Pap. PNW-RP-570. Portland, OR: U.S. Department of Agriculture, Forest Service, Pacific Northwest Research Station. 41 p.

Rai, S. 2009. Land use and climate change. New York: Nova Science Publishers. 131 p.

Richards, K.; Stokes, C. 2004. A review of forest carbon sequestration cost studies: a dozen years of research. Climatic Change. 63: 1–48.

Schlenker, W.; Roberts, M.J. 2008. Estimating the impact of climate change on crop yields: the importance of nonlinear temperature effects. NBER Work. Pap. 13799. Cambridge, MA: National Bureau of Economic Research, Inc. 37 p.

Sohngen, B.; Brown, S. 2004. Measuring leakage from carbon projects in open economies: a stop timber harvesting project as a case study. Canadian Journal of Forest Research. 34: 829–839.

Sohngen, B.; Mendelsohn, R. 1998. Valuing the market impact of large scale ecological change: the effect of climate change on U.S. timber. American Economic Review. 88(4): 689–710.

Sohngen, B.; Mendelsohn, R. 2003. An optimal control model of forest carbon sequestration. American Journal of Agricultural Economics. 85(2): 448–457.

Struck, D. 2010. Buying carbon offsets may ease eco-guilt but not global warming. Christian Science Monitor. April 20. (April 21, 2010).

U.S. Department of Agriculture, Economic Research Service [USDA ERS]. 2006. Agricultural resources and environmental indicators. Washington, DC. 54 p.

U.S. Department of Agriculture, National Agricultural Statistics Service. [USDA NASS]. 2009. Land values and cash rents: 2008 summary. Washington, DC. 27 p.

U.S. Environmental Protection Agency. [USEPA]. 2010. Climate change indicators in the United States. http://www.epa.gov/climatechange/ indicators.html. (April 28, 2010).

van der Vink, G.; Allen, R.M.; Chapin, J.; Crooks, M.; Fraley, W.; Krantz, J.; Lavigne, A.M.; LeCuyer, A.; MacColl, E.K.; Morgan, W.J.; Ries, B.; Robinson, E.; Rodriquez, K.; Smith, M.; Sponberg, K. 1998. Why the United States is becoming more vulnerable to natural disasters. Eos, Transactions, American Geophysical Union. 79(44): 533–537.

Wear, D.N.; Newman, D.H. 2004. The speculative shadow over timberland values in the U.S. South. Journal of Forestry. 102(8): 32–38.

White, E.; Alig, R.J.; Haight, R.G. 2010. The forest sector in a climate-changed environment. In: Alig, R.J., tech. coord Economic modeling of effects of climate change on the forest sector and mitigation options: a compendium of briefing papers. Gen. Tech. Rep. PNW-GTR-833. Portland, OR: U.S. Department of Agriculture, Forest Service, Pacific Northwest Research Station: 1–35. Chapter 1.

White, E.; Morzillo, A.T.; Alig, R.J. 2009. Land development indicators and projections of rural land conversion under different scenarios. Landscape and Urban Planning. 89: 37–48.

Wu, J.; Lin, H. 2010. The effect of the conservation reserve program on land values. Land Economics. 86(1): 1–21.

Chapter 3: Socioeconomic Impacts of Climate Change on Rural Communities in the United States

Pankaj Lal, Janaki Alavalapati, and D. Evan Mercer

Introduction

Climate change refers to any distinct change in measures of climate such as temperature, rainfall, snow, or wind patterns lasting for decades or longer (USEPA 2009). In the last decade, there has been a clear consensus among scientists that the world is experiencing a rapid global climate change, much of it attributable to anthropogenic activities. The extent of climate change effects (e.g., future temperature increase) is difficult to project with certainty, as scientific knowledge of the processes is incomplete and the socioeconomic factors that will influence the magnitude of such increases are difficult to predict (IPCC 2001). However, even if greenhouse gas (GHG) emissions are reduced significantly over the coming years, significant increases in temperature and sea level rise may still occur.

The impacts of climate change can be broadly grouped under three headings: ecological, social, and economic. The ecological impacts of climate change include shifts of vegetation types and associated impacts on biodiversity; change in forest density and agricultural production; expansion of arid land; decline in water quantity and quality; and stresses from pests, diseases, and wildfire. Salient social impacts may include changes in employment, equity, risk distribution, and human health, and relocations of populations. Economic impacts include increased risk and uncertainty of forest or agricultural production, alteration in productivity for crops and forest products, reduction in supply of ecosystem goods and services, increased cost of utilities and services, and altered energy needs.

Climate change will most likely affect populations through impacts on the necessities and comforts of life such as water, energy, housing, transportation, food, natural ecosystems, and health systems. Considerable uncertainty remains about the nature and magnitude of climate change impacts, particularly those related to rural communities, in view of (1) the complex nature of farm decisionmaking, in which there are many nonclimatic issues to manage; (2) the likely diversity of responses within and between regions; (3) the difficulties that might arise if climate changes are nonlinear or increase climate extremes; (4) timelags in responses of communities; and (5) the possible interactions among different adaptation options and economic, institutional, and cultural barriers that inhibit such strategies. In light of these uncertainties, there is a need to increase our understanding of how ecosystems, social and economic systems, human health, and infrastructure will be affected by climate change in the context of other stresses.

The climate change literature specifically addressing social and economic effects of climate change in rural versus urban areas is limited. Although potential threats to urban and rural populations have been described in recent reports (e.g., USGCRP 2009), information delineating the impacts of climate change specifically on rural communities is scarce. The research has largely been sector-specific, such as delineating impacts on agriculture, health, transportation, demography, energy, etc., and has rarely addressed how these impacts might differ across urban or rural communities.[1] Similarly, knowledge of the comparative impacts of climate change in different geographical regions is limited. Because much of the climate change literature does not specifically address social and economic effects of climate change, we make inferences about these effects from national or sector assessments dealing largely with biophysical impacts. In addition, very few studies have attempted to delineate impacts across different spatial scales in terms of severity. Also, it is difficult to compare severity of impacts in light of future uncertainties. The capacity of the community to act in response to climate change and community resilience has been largely absent in climate change research (Flint and Luloff 2005). However, delineating the impacts of climate change on rural populations is critical, as they tend to depend on climate-sensitive livelihoods and are especially vulnerable to climate change events.

One difficulty in analyzing the impacts of climate change on rural communities, is the lack of a clear demarcation between rural and urban areas, as evidenced by the wide variety of definitions of "rural" employed by researchers and policymakers. For example, the USDA Economic Research Service (USDA ERS 2010g) lists as many as nine definitions for "rural." Whether an area is categorized as rural or urban depends in large part on how urban spaces are demarcated, i.e., whether urban spaces are defined in terms of administrative boundaries, land use patterns, or economic influence, and the minimum population thresholds established for delineating areas as urban or rural (Cromartie and Bucholtz 2008). Administrative definitions identify urban space along municipal or other jurisdictional boundaries. Definitions based on land use demarcate urban areas based on population density, whereas economic definitions incorporate the influence of cities beyond densely settled cores and demarcate based on broader commuting areas. The Office of Management and Budget (OMB) identifies counties as rural if they are not core counties (core counties contain one or more urban areas of 50,000 people or more) or economically tied to the core counties, as measured by the share of the employed population that commutes to and from core counties. For the purpose of this study,

---

[1] The literature that focuses on indigenous communities tends to be more developed than the literature on rural areas in general.

we follow the OMB definition and discuss impacts of climate change on nonmetro (rural) areas comprising about 2,052 counties lying outside metro boundaries.

Vulnerability of Rural Communities

Rural regions contain about 17 percent of the U.S. population but extend across 80 percent of the land area (fig. 3-1). The communities residing in these areas differ from their urban counterparts in terms of demography, occupations, earnings, literacy, poverty incidence, dependency on government funds, housing stress, mortality rates, etc. These differences tend to reshape economic and sociocultural conditions across rural counties and can provide insights as to why rural populations might have different vulnerabilities[2] to climate change than their urban counterparts. Vulnerability is a function of the character, magnitude, and variability rate of climate change to which a community is exposed, and the community's sensitivity and adaptive capacity (IPCC 2007). The community adjusts (adapts) in response to actual or expected climatic stimuli or their effects, in order to mitigate (moderate) adverse impacts or exploit beneficial opportunities. The higher a community's adaptive capacity, the lower is its vulnerability to climate change.

---

[2] Vulnerability is defined as the degree to which a system is susceptible to, and unable to cope with, adverse effects of climate change, including climate variability and extremes (IPCC 2007).

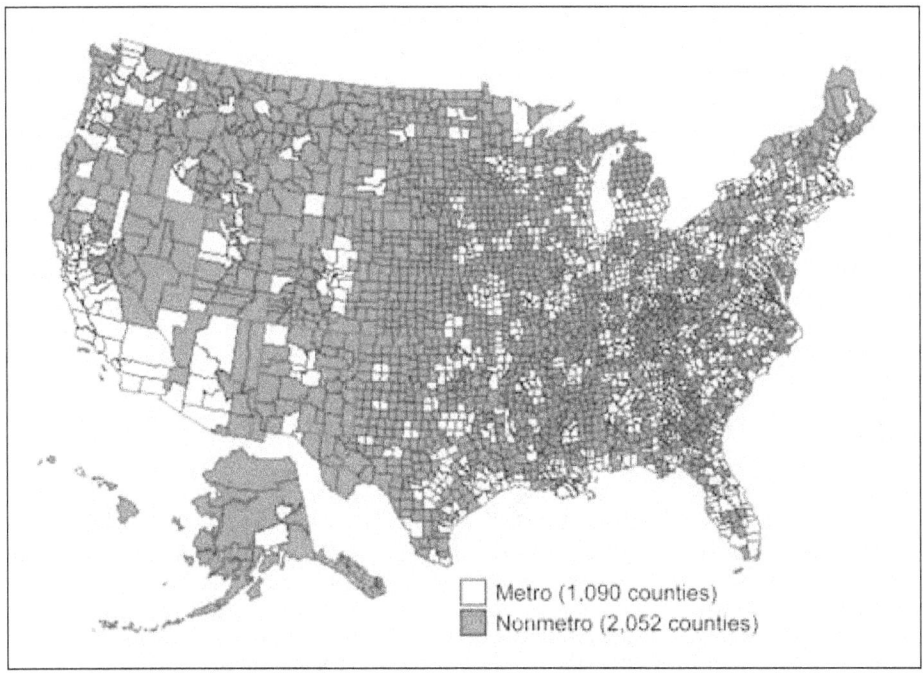

Figure 3-1—Nonmetro and metro counties, 2003. Source: USDA ERS 2010f.

Because 19.6 percent of nonmetro countries are farm dependent[3] as compared to just 3.4 percent of metro counties (USDA ERS 2010d) (fig. 3-2), rural communities are expected to disproportionately experience the brunt of the climatic impacts on agriculture. However, the specific impacts will vary across the United States. For example, the Midwest and Great Plains regions where farming is the predominant land use should experience larger impacts compared to other regions such as the Southeast, Northeast, or Lake States.

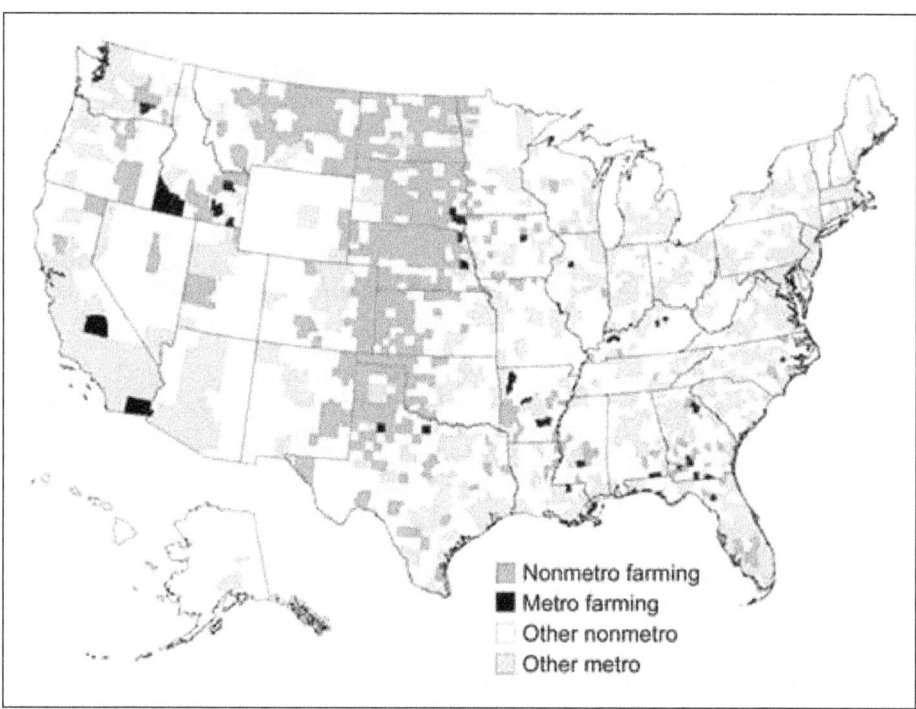

Figure 3-2—Farming-dependent counties, 1998–2000. Source: USDA ERS 2010c.

Rural counties tend to be poorer than their urban counterparts, as shown in figure 3-3. The per capita income in urban areas ($32,077) far exceeded per capita income in micropolitans ($23,338) (counties with cities of 10,000 to 49,999 residents and socioeconomic ties to adjacent counties), and noncore counties with neither a city over 10,000 residents nor socioeconomic ties to a city of that size, ($21,005) (USDA ERS 2010j). Among rural counties, per capita income is generally higher in the Northeast than in the Southeast or Southwest. The lower rural earning levels indicate lower shares of highly skilled jobs and lower returns to college degrees in rural labor markets (USDA ERS 2010l). Unemployment is also often higher in rural

---

[3] Farming-dependent counties have either (1) 15 percent or more of average annual labor and proprietors' earnings derived from farming or (2) 15 percent or more of employed residents who work in farm occupations. See USDA ERS 2010e for details.

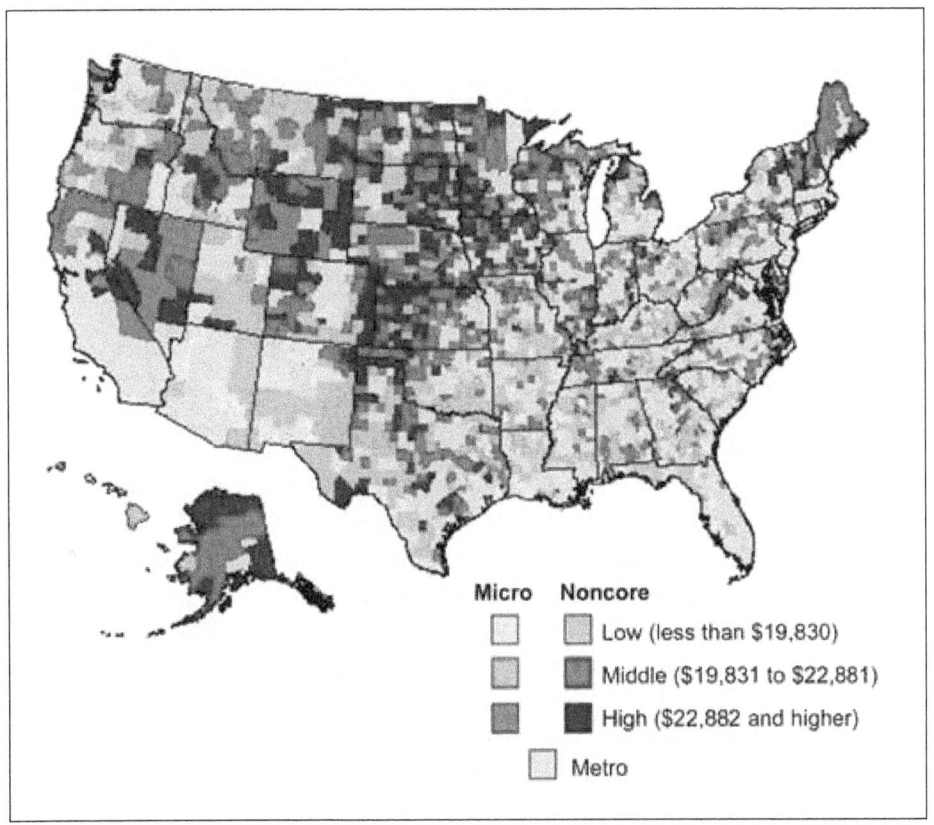

Figure 3-3—Per capita income in micropolitan and noncore counties, 2001. Source: USDA ERS 2010j.

areas. For example, 396 of the 460 counties classified as having low employment[4] were rural (Whitener and Parker 2007). The rural regions in the Southeast stand out as being plagued by high unemployment. Higher unemployment suggests a higher sensitivity and lower capacity to cope with the adverse impacts of climate change.

The rural-urban income gap has been widening recently (USDA ERS 2010j). For example, between 1993 and 2004, rural areas averaged 0.5 percent annual growth in real earnings compared to 1.2 percent per year in urban areas (USDA ERS 2010h). The rural-urban income gap is associated with lower costs of living in rural areas, lower educational attainment, less competition for workers among employers, and fewer highly skilled jobs in the rural occupational mix. As vulnerability to climate change is directly related to income levels (Yohe and Tol 2002), the rural communities' vulnerability to climate-related risk is expected to be higher than that of urban communities. Incidence of poverty is another factor that will influence a community's vulnerability to climate change. Turner et al. (2003)

---

[4] Less than 65 percent of residents 21 to 64 years old were employed in 2000. See USDA ERS 2010e for details.

suggested that the poor and marginalized in Canada and the United States have historically been most at risk from weather shocks. Rural communities also tend to suffer from higher incidence of persistent poverty as evidenced by the data for counties[5] (fig. 3-4).

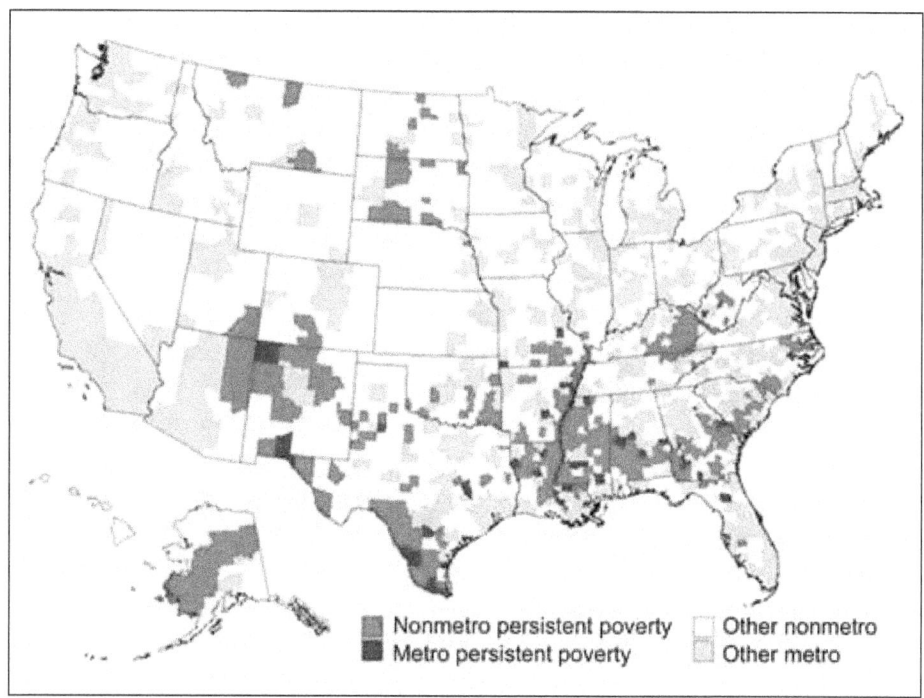

Figure 3-4—Persistent poverty counties, 1970–2000. Source: USDA ERS 2010i.

Rural communities tend to be less ethnically diverse than urban areas (USDA ERS 2010i). Of the 442 rural counties categorized as high-poverty counties in 2000, three-fourths were classified as predominately Black, Hispanic, or Native American counties. There were 210 predominantly Black high-poverty counties, mostly in the Southeast; 72 Hispanic high-poverty counties, mostly in Texas and New Mexico; and 40 high-poverty counties dominated by Native Americans primarily in Alaska, New Mexico, Utah, Arizona, Oklahoma, Montana, South Dakota, and North Dakota. Of the remaining one-fourth of high-poverty counties, most (91 counties) are in the southern highlands of eastern Kentucky, West Virginia, and parts of Missouri and Oklahoma and are dominated by non-Hispanic Whites. The remainder of the high-poverty counties (27) includes thinly populated farming areas in the northern Great Plains, where annual income levels range widely depending on

---

[5] Counties in which 20 percent or more residents were poor as measured by last four censuses, 1970–2000. See USDA ERS (2010e) for details.

wheat and cattle prices and output, and two high-poverty counties where Asians are the dominant ethnic group.

The Intergovernmental Panel for Climate Change (IPCC) identifies economic wealth, technology, information and skills, infrastructure, institutions, and equity as significant features of adaptive capacity (Smit et al. 2001). Wealthier communities tend to have greater access to technology, information, developed infrastructure, and stable institutions (Easterling et al. 2007) and thus possess higher adaptive capacity for climate change. According to the USDA Economic Research Service (2010i), rural communities in the South and West account for approximately 59 percent of the total rural population in the country and have the highest poverty rates in the country. Thus, we would expect these areas to have generally lower adaptive capacity to cope with future climate risks (fig. 3-5). Just because a community may have high socioeconomic status, however, does not mean it is effective at making collective decisions and meeting the needs of the broader population. Social relations are difficult to quantify and compare across communities and regions. In this paper, we primarily use socioeconomic status, technology, infrastructure, and skills to make inferences about the relative adaptive capacity of communities.

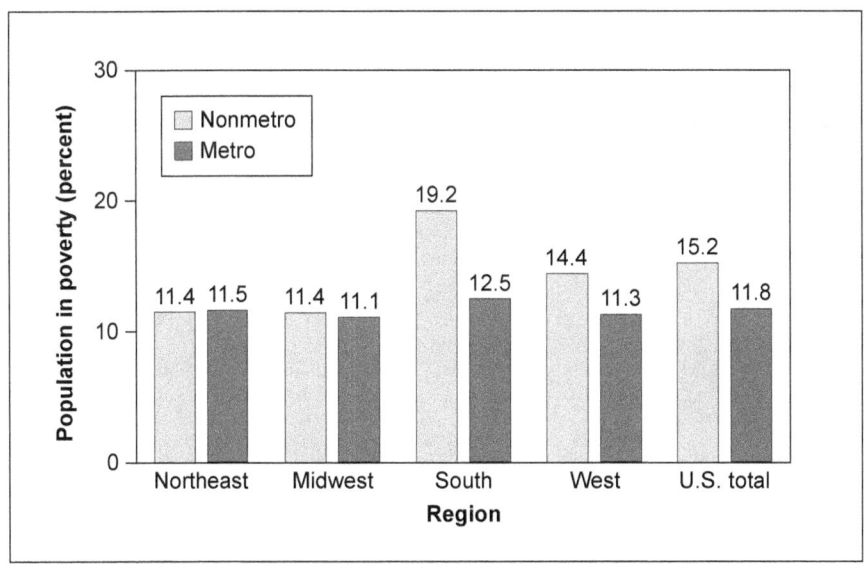

Figure 3-5—Poverty rates by region and metro status, 2006. Source: USDA ERS 2010b.

Another factor that adds to the vulnerability of rural areas is their dependence on government transfer payments. Based on 2001 data (USDA ERS 2010k), government transfer payments averaged $4,365 per person per year in rural, nonmetro areas compared to $3,798 for metro areas. Federal and state government transfers accounted for about one-fifth of rural income as compared to just one-eighth of metro income. Unless government transfer payments to rural areas are able to keep up with increased need resulting from climate change impacts, rural areas may experience greater vulnerability to climate change than urban areas.

Most outdoor recreation areas are in rural counties of the United States; for example, 334 of the 368 (91 percent) recreation-dominated counties[6] were rural and only 34 were urban in 2003 (Whitener and Parker 2007) (fig. 3-6). Many of the jobs that are usually associated with recreation, such as those in hotels and restaurants, often are low paying with few fringe benefits (Deller et al. 2001). However, in rural areas, which have lower incomes and higher poverty incidence than their urban counterparts, even these low-paying jobs might be quite important for the livelihoods of communities. If climate change dramatically reduces or shifts job opportunities in recreation, most of the impact will be felt by rural communities, where most of the recreation employees reside (Morello et al. 2009). However, as

---

[6] Rural counties have been classified using a combination of factors by Economic Research Service, including share of employment or share of earnings in recreation-related industries in 1999, share of seasonal or occasional-use housing units in 2000, and per capita receipts from motels and hotels in 1997. See USDA ERS 2010e for details.

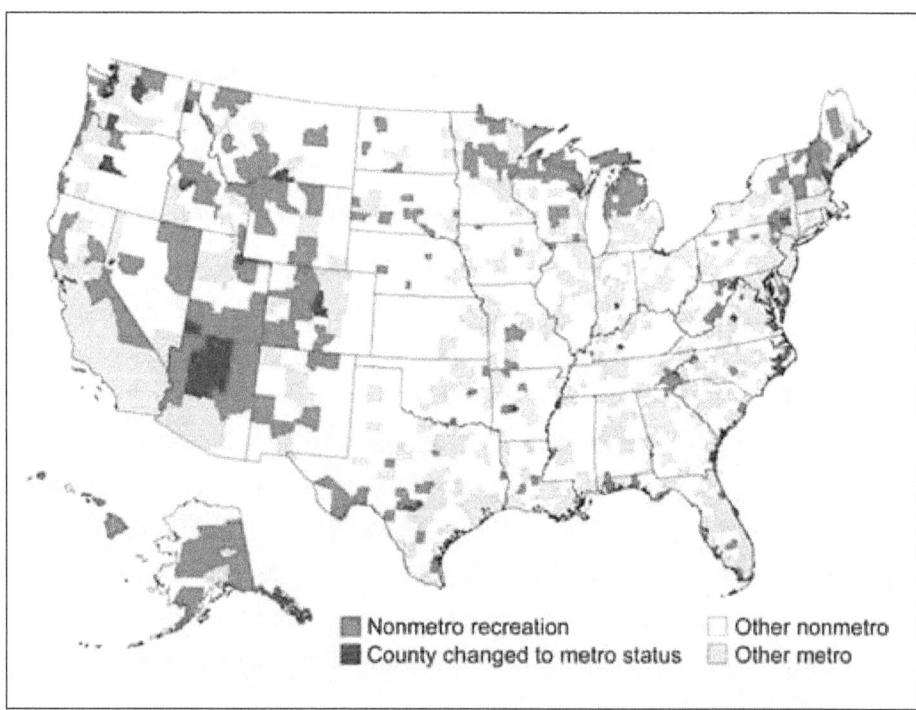

Figure 3-6—Nonmetro recreation counties, 1999. Source: Source: USDA ERS 2010a.

we discuss below, impacts of climate change on recreation will vary considerably geographically.

Rural residents tend to have higher rates of age-adjusted mortality, disability, and chronic disease than their urban counterparts, although mortality and disability rates vary more by region than by metro status (Jones et al. 2009). Furthermore, as young adults move out of small, rural communities, many rural communities tend to reflect an increasingly vulnerable demographic of very old and very young people, placing them more at risk for climate change effects than urban communities. Climate impacts, coupled with demographic shifts in rural communities, may make it more difficult to supply adequate and efficient public health services and educational opportunities to rural areas (USGCRP 2009). Detrimental climate change effects are also likely to be compounded by additional stresses and disturbances such as increased land use change, pollution, wildfires, and invasive species (USGCRP 2009).

The accessibility of health care resources tends to decline as population density declines and geographic isolation increases. As a result, rural residents tend to face higher financial and travel costs to access health care and pay a greater share of household income for health care than their urban counterparts (Jones et al. 2009). Furthermore, emergency response systems are often less effective in rural areas because the population is dispersed and geographically isolated. The combined

Ralph Alig

Climate change may have different effects geographically and on different parts of rural and urban communities.

effects of changing demographics and increasing health costs are likely to make it more difficult to supply adequate and efficient public health services to rural areas in the future. Therefore, with lower access to health infrastructure and higher proportion of income spent on health services, rural communities are likely to be more vulnerable to adverse health impacts caused by climate change.

A changing climate will mean reduced opportunities for some activities and locations and expanded opportunities for others, leading to regional differentiation of impacts in term of incidence and intensity. Certain facets of climate change may impact one particular region but not others. For example, increased risk of drought, pests, and extreme weather events may add additional economic stress and tension to rural communities (Motha and Baier 2005, Parton et al. 2007). However, as we explore below, climate-driven shifts in crop types and recreation areas may also benefit some rural communities. Similarly, impacts on water quality and quantity owing to climate change will differ across regions. For example, reduced snowpack and earlier snowmelt because of warmer temperatures will alter the timing and amount of water supplies, impacting the Western United States harder than the Eastern United States (USGCRP 2009). Climate change events may also differentially affect the culture and livelihood patterns of indigenous communities in the United States.

Impacts on Rural Communities

Climate change impacts will differ by region and sector, and so will the capacity to handle the resulting challenges. Climate change will likely produce a range of impacts depending on specific attributes of the affected rural communities or industries. Some communities may benefit from climate-induced changes, whereas others may face large losses. For example, communities dependent on oil and gas extraction and mining-related industries are likely to experience climate change differently than predominantly agricultural communities (USGCRP 2009). Furthermore, agricultural communities in different parts of the United States will likely also differ in how climate change affects them. For example, farming communities in the Great Lakes States may benefit from warming climate owing to improved growing conditions for some crops (like fruit production) that are currently limited by length of growing season and temperature, whereas farming communities in the Midwest may face adverse impacts of climate change owing to lower availability of irrigation water (Hatfield et al. 2008).

Social Impacts

Important characteristics of rural society make it vulnerable to climate change impacts and affect how the risks and costs may be distributed among different regions. Salient social impacts include impacts on human health via direct effects (e.g., thermal stress) and indirect effects (e.g., disease vectors and infectious agents), increase in societal conflicts, and high vulnerability of particular community groups such as Native Americans.

Human health—

Climate change will affect human health through both direct and indirect pathways. Direct impacts will result from increased exposure to temperature (heat waves, winter cold) and other extreme weather events (floods, cyclones, storm surges, droughts) and increased production of air pollutants and aeroallergens such as spores and molds (USGCRP 2009). Figure 3-7 shows temperature changes projected under two GHG emission scenarios. The average U.S. temperature is projected to increase by approximately 7 to 11 °F for the higher emissions scenario and 4 to 6.5 °F for the lower emission scenario (USGCRP 2009). Although most of the country will face greater warming in summer than in winter, Alaska is expected to experience far more warming in winter than summer (Christensen et al. 2007).

The occurrence of extreme heat events like the Chicago heat wave of 1995, which lasted for 5 days and resulted in an 85-percent increase in heat-related mortality and an 11-percent increase in heat-related hospitalizations, are expected to become more frequent as a result of climate change. However, rural counties, which have lower builtup area than cities, should be less vulnerable to extreme heat events. This is due to the fact that concrete and asphalt in cities absorb and hold heat, while tall buildings prevent heat from dissipating and reduce air flow leading to a "heat island effect." The larger amounts of vegetation in rural areas also tend to provide more shade and evaporative cooling than in urban areas.

Human health may also be indirectly affected by an increase in water, food, and vector-borne diseases. Kilpatrick et al. (2008) suggested that increasing temperatures significantly increases dissemination and transmission of viral infection, most likely through increased viral replication. The distribution and abundance of vector organisms and intermediate hosts can also be affected by physical factors such as temperature, precipitation, humidity, surface water, and wind and by biotic factors such as vegetation, host species, predators, competitors, and parasites, all of which may be altered by climate change (PSR 2010).

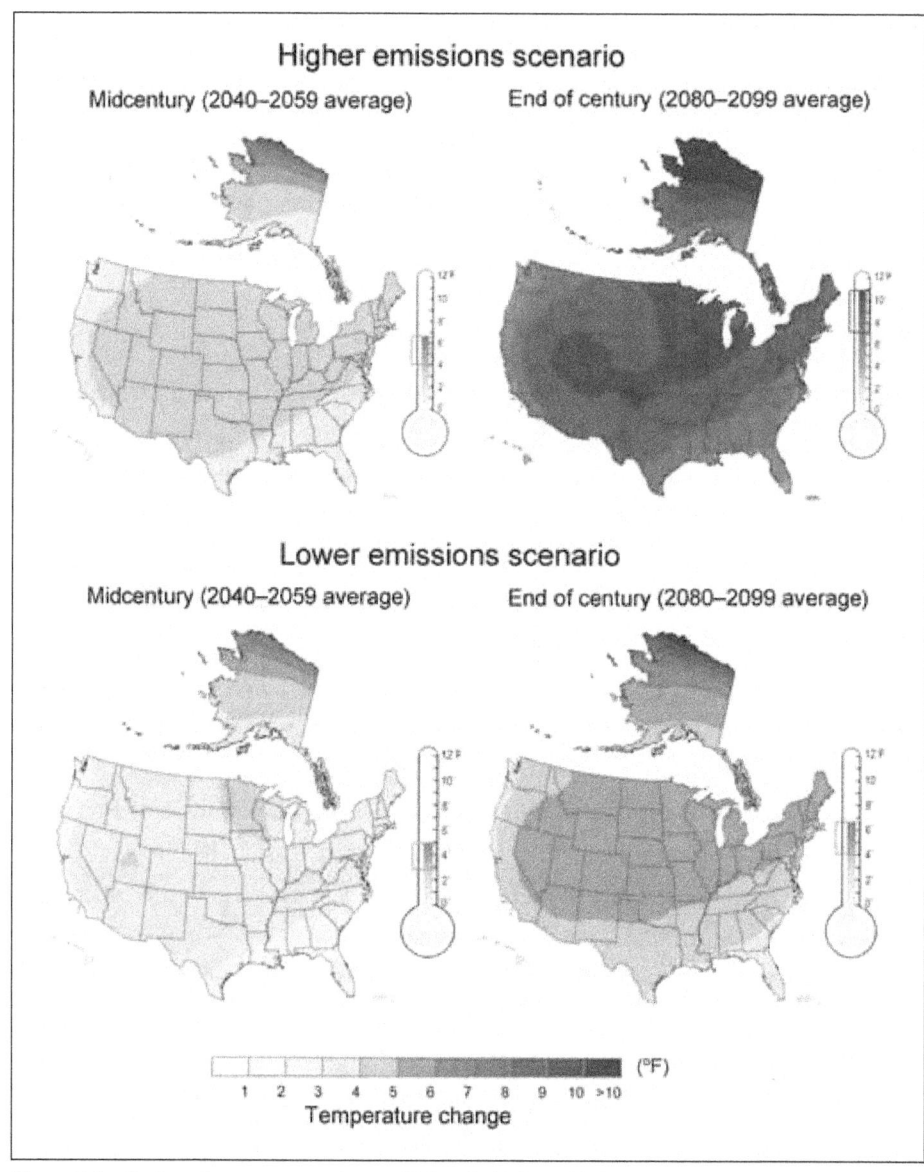

Figure 3-7—Projected temperature change (Fahrenheit) from 1961–1979 baseline for two emission scenarios. Based on projections of future temperature by 16 of the Coupled Model Intercomparison Project Three climate models for IPCC higher and lower scenarios. The brackets on the thermometers represent the likely range of model projections, although lower or higher outcomes are possible. Source: USGCRP 2009.

There are clear trends of increasing incidents of very heavy precipitation in the Nation as a whole, and particularly in the Northeast, Midwest, Alaska, and islands as shown in figure 3-8.

Heavy downpours can lead to increased sediment in runoff and outbreaks of waterborne diseases (Ebi et al. 2008, Field et al. 2007). Degradation of water quality and increases in pollution carried to lakes, estuaries, and the coastal ocean

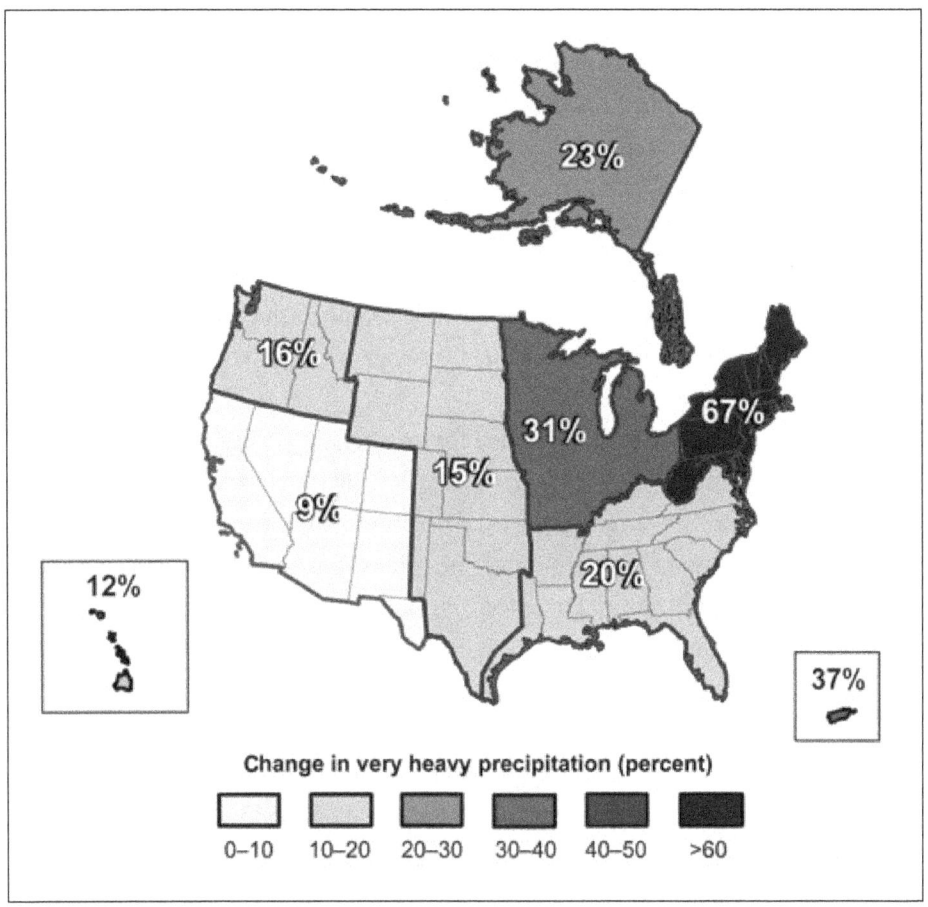

Figure 3-8—Increases in amounts of very heavy precipitation, 1958–2007. Source: USGCRP 2009 based on Groisman et al. 2004.

following heavy downpours, especially when coupled with increased temperature, can also result in blooms of harmful algae and bacteria and increased risk of waterborne parasites such as *Cryptosporidium* and *Giardia*. Incidences of heavy rain and flooding can also contaminate food crops with feces from nearby livestock or wild animals, increasing the likelihood of food-borne disease (Ebi et al. 2008). Projected increases in carbon dioxide (CO_2) can also stimulate the growth of stinging nettle and leafy spurge, two weeds that cause rashes when they come into contact with human skin (Ziska 2003).

Impacts on indigenous communities—

Native American communities, which are predominantly rural, may face disproportionately higher levels of climate change impacts on their livelihoods, rights and access to natural resources, future growth, and in some cases, their culture, which depends on traditional ways of collecting and sharing food

(Hanna 2007, Nilsson 2008, Tsosie 2007, USGCRP 2009). For many indigenous communities, climate change may also reduce the availability and accessibility of such traditional food sources as seals, whose migration patterns depend on their ability to cross frozen rivers and wetlands (USGCRP 2009). It is estimated that climate change may increase flooding and erosion in 184 out of 213, or 86 percent, of Native Alaskan communities (USGAO 2003). Native cultures in the Great Plains and Southwest are also vulnerable to climate change effects. Many of these tribes have limited capacity to respond to climate change and already face severe problems with water quantity and quality—problems likely to be exacerbated by climate change.

Relocation options tend to be limited for many Native Americans who live on established reservations and may be restricted to reservation boundaries (NAST 2001). Having already been relocated to reservations, these communities have historically been disconnected from their traditional life, prohibited from engaging in important social and cultural practices, and allowed limited participation in land management and planning (Tsosie 2007). Furthermore, Native American communities may be more vulnerable to climate change impacts, as their rights and livelihoods tend to be interwoven with specific lands limiting their relocation options in the face of alterations in resource availability (Donoghue et al. 2009).

The melting of permafrost, which has already turned solid ground into mush in some places in Alaska, threatens the economies and cultures of many Alaskan tribes as they may be required to relocate at large economic and cultural cost (NTAA 2009). For example, the way of life of the Inupiaq Tribe in Alaska is threatened owing to climate change. The traditional method of food storage of Inupiaqs is being disrupted by warming, as "permafrost" does not remain permanent, leading their belowground storages (*sigulaqs* in native language) to be thawed and sometimes flooded with meltwater. The resulting spoiled meat increases the risk of food-related illness.

Subsistence cultures such as Native Alaskans adapt to year-to-year fluctuations of game species by shifting practices and target species, which implies some ability to adapt to effects of near-term climate change (USGCRP 2009). Integrating the adaptation insights of indigenous peoples in terms of access, process, and the outcomes of climate policy and planning should be helpful in reducing impacts (Nilsson 2008). However, these adaptation opportunities may be severely constrained, as warming in Alaska is especially likely to reshape patterns of human settlement and intertwined economic activities (Wilbanks et al. 2007).

Economic Impacts

Major parts of the rural economies of the United States are directly sensitive to climate change, including the agriculture, recreation and tourism, forestry, water resources, energy, and fisheries sectors.

Agriculture—

Agriculture will certainly face significant changes from climate change. Longer growing seasons and increased CO_2 have positive effects on some crop yields, although this might be counterbalanced in part by the negative effects of additional disease-causing pathogens, insect pests, and weeds (USGCRP 2009). Hatfield et al. (2008) suggested that even moderate increases in temperature will decrease yields of corn, wheat, sorghum, bean, rice, cotton, and peanut crops. More frequent temperature extremes will also create problems for crops. For example, tomatoes, which are well-adapted to warmth, produce lower yields or quality when daytime maximum temperatures exceed 90 °F for even short periods during critical reproductive stages (Kunkel et al. 2008). Some crops, however, benefit from higher temperatures, and global warming will likely result in a longer growing season for crops like melon, okra, and sweet potato (Hatfield et al. 2008). Significant technological progress might also temper adverse climate change impacts. For example, corn yields have shown an upward trend even in light of variation caused by climate events (fig. 3-9). However, U.S. Global Change Research Program (2009) argued that it is difficult to maintain this historical upward trend without dramatic technological innovations.

Climate change may increase agricultural production costs for a number of reasons. For example, the expansion of weeds may be exacerbated by climate change if weeds benefit more from higher temperatures and CO_2 levels than traditional crops (Hatfield et al. 2008). With continued warming, invasive weed species are expected to expand northward and increase costs and crop losses as evidenced by the fact that loss of crops owing to weeds is higher in the South than in the North. For example, southern farmers lose 64 percent of the soybean crop to weeds, whereas northern farmers lose just 22 percent (Ryan et al. 2008). Controlling weeds currently costs the United States more than $11 billion a year, with the majority spent on herbicides (Kiely et al. 2004). This cost is likely to increase as temperatures rise. The problem is aggravated by the fact that the most widely used herbicide in the country, *glyphosate*, loses its efficacy at CO_2 levels that are projected to occur in the coming decades (Wolfe et al. 2007). Another potential impact of climate change is premature plant development and blooming, resulting in exposure of young plants and plant tissues to late-season frosts. For

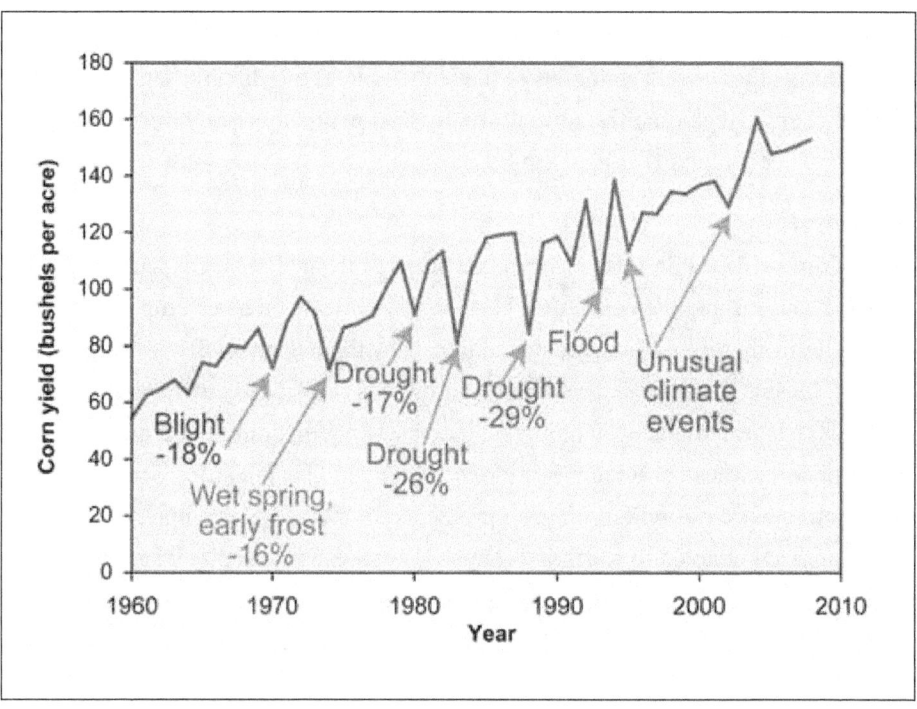

Figure 3-9—United States corn yield trend since 1960. Source: USGCRP 2009 based on NAST 2001.

example, the 2007 spring freeze in the Eastern United States led to widespread devastation of crops and natural vegetation, because the frost occurred during the flowering period of many trees and during early grain development on wheat plants (Gu et al. 2008).

Climate change is projected to increase the intensity of precipitation, resulting in heavy downpours across the country (Kunkel et al. 2008). This excessive rainfall may delay spring planting, in turn lowering profits for farmers who are paid a premium for early-season production of high-value crops such as melons, sweet corn, and tomatoes. Field flooding during the growing season by heavy downpours leads to low oxygen levels in the soil, higher susceptibility to root diseases, and increased soil compaction from the use of heavy farm equipment on wet soils. Increased intensity of precipitation can also result in reduced quality of many crops.

The projected warmer temperatures are expected to increase livestock production costs owing to lower feed intake and increased requirements for energy to maintain healthy livestock at higher temperatures. Forage production may also be affected by climate change. Rising atmospheric CO_2 concentrations can increase the quantity of forage produced, but it might reduce forage quality, as plant nitrogen and protein concentrations often decline with higher concentrations of CO_2 (Hatfield et al. 2008). The dairy industry is also quite sensitive to temperature changes, as dairy cows' productivity decreases above 77 °F (25 °C). By late in this century, all Northeastern States except the northern parts of Maine, New Hampshire, New York, and Vermont are projected to suffer declines in July milk production under the higher emissions scenario (USGCRP 2009). In California, an annual loss of $287 to $902 million is expected for this $4.1 billion industry. In parts of Connecticut, Massachusetts, New Jersey, New York, and Pennsylvania, climate change is projected to produce a large decline in milk production from 10 to 20 percent or greater (USGCRP 2009).

Climate change impacts on rural communities engaged in agriculture will differ across regions; some will likely benefit while others lose depending on their geographic location and adaptive capacity. Heat and water stress from droughts and heat waves is likely to decrease yields and adversely affect crops like wheat, hay, corn, barley, cattle, and cotton in the Great Plains (Motha and Baier 2005). Much of the Northwest region's agriculture will experience detrimental impacts. Particularly impacted will be specialty crops in California such as apricots, almonds, artichokes, figs, kiwis, olives, and walnuts (Lobbel et al. 2006). By late in this century, winter temperatures in many important fruit-producing regions such as the Northeast may be too warm to support fruit production. For example, Massachusetts and New Jersey supply nearly half the Nation's cranberry crop. By the middle of this century, these areas may not be able to support cranberry production owing to lack of winter chilling (Frumhoff et al. 2007, Wolfe et al. 2007).

Emily Jederlinich

Climate change may affect costs and production levels for forestry and agricultural enterprises.

In contrast, warming is expected to improve the climate for fruit production in regions such as the Great Lakes (Field et al. 2007) or Midwest (USGCRP 2009). However, even farms and regions that temporarily benefit from altered environmental conditions (e.g., carbon fertilization and extended growing season) risk economic losses if temperatures exceed those preferred by the crops they currently produce (Ruth et al. 2007). In the Midwest, the projected increases in winter and spring precipitation and flooding are likely to delay planting and crop establishment.

Recreation and tourism—
Outdoor recreation activities depend on the availability and quality of natural resources, such as beaches, forests, wetlands, snow, and wildlife (USGCRP 2009). Johnson and Beale (2002) identified 329 recreation-dependent counties by geographic location, natural amenities, and form of recreation.[7] Most of the rural recreation counties are concentrated in the West, the Upper Great Lakes, and the Northeast regions (Reeder and Brown 2005). Recreation counties in general tend to have relatively low population densities, and more of their residents live in rural parts of the county (Jones et al. 2009). In the West, rural counties reflect opportunities for hiking, mountain climbing, fishing, and wintertime sports found in the many national parks and ski resorts. On the other hand, recreation-dependent counties in the Upper Great Lakes and Northeast—especially in New England and Upstate New York—are largely due to second homes in areas with lakes. Many of these areas also have significant wintertime recreation activities, including snowmobiling and skiing.

Increased temperatures and precipitation owing to climate change are expected to have a direct effect on the enjoyment of tourism activities, and on the desired number of visitor days and associated levels of visitor spending and employment. Climate change could affect recreation through three pathways: winter activities such as downhill and cross-country skiing, snowshoeing, and snowmobiling; nature

---

[7] Recreational counties were identified based on a multistep selection procedure combining empirical measures of recreational activity along with a review of recreation-related contextual material existing in the counties. These empirical measures used were (1) wage and salary employment in entertainment and recreation, accommodations, eating and drinking places, and real estate as a percentage of all employment reported in the Census Bureau's County Business Patterns for 1999; (2) percentage of total personal income reported for these same categories by the Bureau of Economic Analysis; (3) percentage of housing units intended for seasonal or occasional use reported in the 2000 Census; and (4) per capita receipts from motels and hotels as reported in the 1997 Census of Business. For details see Johnson and Beale 2002.

tourism and related activities such as biking, walking, hunting, etc.; and water-related sports such as diving, sailing, and fishing. A changing climate will mean reduced opportunities for some activities and locations and expanded opportunities for others (Sussman et al. 2008). The length of the season, and, in many cases, the desirability of popular activities like walking; visiting a beach, lakeshore, or river; sightseeing; swimming; and picnicking might increase because of small near-term increases in temperature. On the other hand, snow- and ice-dependent activities, including skiing, snowmobiling, and ice fishing, could be adversely affected by even small increases in temperature. Hunting and fishing opportunities will change as animals' habitats shift and as relationships among species in natural communities are disrupted by their different responses to rapid climate change (USGCRP 2009). In the longer term, as climate change affects ecosystems and seasonality becomes more pronounced, the net economic effect on recreation and how it will influence different population groups in different regions is not known (Wilbanks et al. 2007).

The impact of climate change on ski, snowmobile, and other winter sport industries is expected to be more pronounced in the Northeast and Southwest regions. The ski resorts in the Northeast have three climate-related criteria to remain viable: the average length of the ski season must be at least 100 days, there must be a good probability of being open during the lucrative winter holiday between Christmas and the New Year, and there must be enough nights that are sufficiently cold to enable snowmaking operations. By these standards, only one area in the region (fig. 3-10) is projected to be able to support viable ski resorts by the end of this century under a higher emissions scenario (USGCRP 2009).

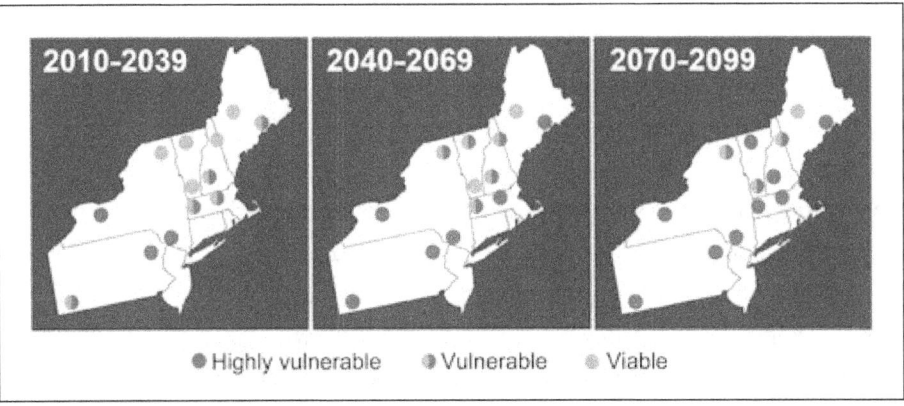

Figure 3-10—Ski areas at risk in the Northeastern United States. Source: USGCRP 2009 based on Frumhoff et al. 2007.

Reduced snowmaking in the Southwest owing to climate change is also expected to shorten the ski season substantially, with projected losses of 3 to 6 weeks (by the 2050s) and 7 to 15 weeks (2080s) in the Sierra Nevada of California (Hayhoe et al. 2004, Scott and Jones 2005). Projections indicate later snow and less snow coverage in ski resort areas, particularly those at lower elevations and in the southern part of the Southwest region. Decreases from 40 to almost 90 percent are likely in end-of-season snowpack under a higher emissions scenario in counties with major ski resorts from New Mexico to California (Zimmerman et al. 2006). The snowmelt dates are also projected to shift earlier at most sites (fig. 3-11).

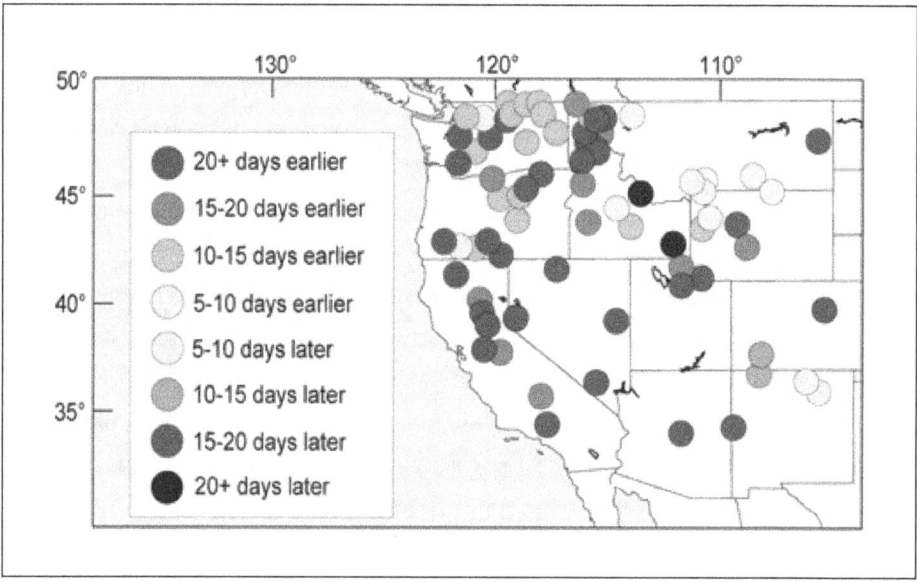

Figure 3-11—Observed spring snowmelt dates in Western United States. Source: USGS 2005.

In addition to shorter seasons, earlier wet snow avalanches could force ski areas to shut down many runs before the season would otherwise end (Lazar and Williams 2008). Resorts require a certain number of days just to break even; cutting the season short by even a few weeks, particularly if those occur during the lucrative holiday season, could easily render a resort unprofitable. The snowmobiling industry is also vulnerable to climate change as it relies on natural snowfall. Some predict that by the 2050s, a reliable snowmobile season will disappear from most regions of the East (Scott and Jones 2006, Scott et al. 2008).

Nature-based tourism—hiking; camping; bird watching; visiting a beach, lakeshore, or river; sightseeing; swimming; and picnicking—is a major market segment in many parts of the country, with over 900 million visitor-days in national, state, and local parks reported in 2001 (USGCRP 2009). The length of the nature tourism season is likely to be enhanced by small near-term increases in temperature. Visits to national parks are projected to increase by 9 to 25 percent (2050s) and 10 to 40 percent (2080s) as a result of a lengthened warm-weather tourism season (Scott and Jones 2006). Nearby communities may benefit economically, but visitor-related ecological pressures could be exacerbated in some parks. Activities like hunting and wildlife-related tourism will change as habitats shift and relationships among species in natural communities are disrupted by their different responses to rapid climate change (USGCRP 2009). Climate-induced environmental changes (e.g., loss of glaciers, altered biodiversity, fire- or insect-impacted forests) may also affect nature tourism, although uncertainty is higher regarding the regional specifics and magnitude of these impacts (Richardson and Loomis 2004, Scott et al. 2007).

The impacts on water-related tourism are likely to be exacerbated by rising sea levels and storm severity especially in areas projected to get drier, such as the Southwest, and in beach communities that are expected to see rising sea levels (Clark et al. 2008, Kleinosky et al. 2005, Williams et al. 2009, Wu et al. 2002). There is evidence that the global sea level is currently rising at an increased rate (Bindoff et al. 2007, Rahmstorf et al. 2007, Vermeer and Rahmstorf 2009). Water sports that depend on the flows of rivers and sufficient water in lakes and reservoirs are already being affected, and much larger changes are expected in the future (Sussman et al. 2008). Higher sea levels may erode beaches, and along with increasing water temperatures, destroy or degrade natural resources such as mangroves and coral reef ecosystems that attract tourists (Mimura et al. 2007). However, the vulnerability of key recreation areas in the coastal United States to climate change events has not been comprehensively assessed (USGCRP 2009).

Recreational fisheries in many rural counties will also be impacted by climate change. For example, approximately half of the wild trout populations are expected to disappear from the Southern Appalachian Mountains owing to rising stream temperatures. Losses of western trout populations may exceed 60 percent in certain regions. About 90 percent of bull trout (*Salvelinus confluentus*), which live in western rivers, are projected to be lost on account of warming. The state of Pennsylvania is predicted to lose 50 percent of its trout habitat in the coming decades, and warmer states such as North Carolina and Virginia, may lose up to 90 percent (Willliams et al. 2007).

The U.S. islands (Hawaii, Puerto Rico, Virgin Islands, Guam, American Samoa) face potentially large impacts from climate change. For island fisheries sustained by healthy coral reefs and marine ecosystems, climate change impacts exacerbate stresses such as overfishing (Mimura et al. 2007), affecting both fisheries and tourism. Any adverse impacts on tourism threaten the livelihood of many island communities. For example, in 1999, the Caribbean Islands had tourism-based gross earnings of $17 billion, providing 900,000 jobs and making the Caribbean one of the most tourism-dependent regions in the world (Heileman et al. 2004).

Forestry—

The impacts of climate change on forestry are expected to arise from shifts in forest distribution and types, increased wildfire risk, increased chance of pest attacks and diseases, and adverse impacts on biodiversity. Projected changes in climate and the consequent impact on forests could affect market incentives for investing in intensive forest management (such as planting, thinning, genetic conservation, and tree improvement) and developing and investing in wood-conserving technologies. The effect on rural communities will differ depending on the geography, demographics, and social and economic conditions each community faces; as with agriculture, some may benefit while others lose.

Potential habitats for trees favored by cool environments are likely to shift north (NAST 2001) (fig. 3-12). As tree species migrate northward or to higher elevations, habitats of alpine and subalpine spruce-fir could possibly be eliminated (IPCC 2007). Aspen and eastern birch communities are also likely to contract dramatically in the United States and largely shift into Canada. Potential habitats that could possibly expand in the United States are oak/hickory and oak/pine in the Eastern United States and ponderosa pine and arid woodland communities in the West. The changing forest distribution is already being observed in many areas. For example, in Colorado, aspen (*Populus* sp. Michx.) has advanced into the more cold-tolerant spruce-fir forests over the past 100 years (Elliott and Baker 2004). The northern limit of the lodgepole pine (*Pinus contorta* Douglas ex Louden) range is advancing into the zone previously dominated by the more cold-tolerant black spruce (*Picea mariana* (Mill.) Britton, Sterns & Poggenb.) in the Yukon (Johnstone and Chapin 2003). In addition, many of the economically valuable timber species in the Midwest—aspen, jack pine (*Pinus banksiana* Lamb.), red pine (*Pinus resinosa* Aiton), and white pine (*Pinus strobus* L.)—may be lost owing to global warming (Easterling and Karl 2001). If the forests in the South and Northeast shift to oak and hickory species in lieu of softwoods, the pulp/wood fiber industry could experience large losses (USGCRP 2009), in turn impacting the rural communities who depend on these industries for their livelihoods.

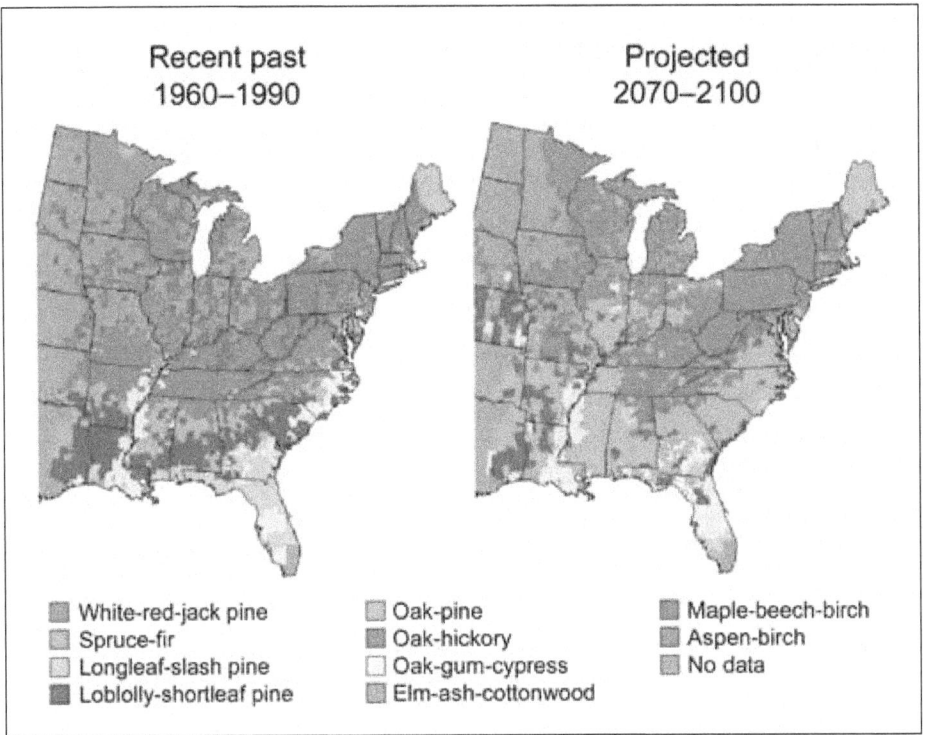

Figure 3-12—Projected shift in forest types in Eastern United States under a midrange warming scenario. Source: USGCRP 2009 based on NAST 2001.

Warmer summer temperatures and reduced rainfall in the West are projected to extend the annual window of wildfire risk by 10 to 30 percent (Brown et al. 2004). These factors are contributing to an overall increase in the area of forest burned each year in the Pacific Northwest and in the United States as a whole (USDA FS 2000). Westerling et al. (2006) analyzed wildfire trends in the Western United States and found a sixfold increase in the area of forest burned since 1986 compared with the 1970–86 period. The average duration of fires increased from 7.5 to 37.1 days—mostly because of an increase in spring and summer temperatures and earlier thawing of snowpacks. The increased incidences of wildfires could affect communities in a number of ways including loss of forest recreation opportunities and increased costs for fire suppression and recovery. For example, Ruth et al. (2007) predicted that the climate-change-induced warming will mean that the state of Washington will face fire suppression cost increases of over 50 percent by 2020 and over 100 percent by 2040, raising the expenses to $93 million and $124 million, respectively. Because many rural communities reside adjacent to forest or are dependent on forest industries for their livelihood, they tend to be directly affected by these wildfires. These wildfires are adversely impacting indigenous communities as well (NTAA 2009).

Climate change is also likely to result in more disturbances from insects, invasive species, and diseases (Alig et al. 2004, Gan 2004, Logan et al. 2003). For example, Ryan et al. (2008) predicted an increase in the frequency and intensity of mountain pine beetle (*Dendroctonus ponderosae*) and other insect attacks, further increasing fire risk and reducing timber production. Insects, historically controlled by cold winters, more easily survive milder winters and produce larger populations in warmer climates. In a changing climate, populations of some pests such as red fire ants (*Solenopsis invicta* Buren), better adapted to a warmer climate, are projected to increase (Cameron and Scheel 2001, Levia and Frost 2004). Invasive weed species that disperse rapidly are likely to find increased opportunities under climate change. Pests can also impact rural communities and especially Native American communities by reducing the availability of nontimber forest products (NTAA 2009).

Damages to forest resources from pests can be significant. For example, spruce bark beetle (*Dendroctonus rufipennis*) outbreaks in the Kenai Peninsula of Alaska (red areas in fig. 3-13) have led to the loss of over 5 million acres of spruce forests. The recent spread of southern pine beetle (*Dendroctonus frontalis* Zimmermann), attributable, in part, to climate change, has affected sawtimber and pulpwood production in Alabama, Louisiana, Mississippi, Tennessee, Kentucky, and the Carolinas. On average, annual losses have reached over 1 percent of gross state product (Ruth et al. 2007).

Changes in temperature and precipitation affect the composition and diversity of native animals and plants through altering their breeding patterns, water and food supply, and habitat availability (Feng and Hu 2007). Therefore, we expect increased extinction of local populations and loss of biological diversity if climate change outpaces species' ability to shift their ranges and form successful new ecosystems. Residents of Alaska are likely to experience the most disruptive impacts of climate change in the near term, including shifts in the range or abundance of wild species crucial to the livelihoods and well-being of indigenous populations (Houser et al. 2001, Parson et al. 2001).

Higher temperatures, decreased soil moisture, and more frequent fires may stress forest ecosystems and ultimately may lead to a conversion of some forests to savannah and grassland (Burkett et al. 2001). Grassland and plains birds, already besieged by habitat fragmentation, could experience significant shifts and reductions in their ranges (Peterson 2003). Biodiversity impacts of climate change may also alter distribution of prominent game and other bird species (e.g., waterfowl, warblers, perching bird species) in many recreational rural counties. The conversion of forest land, habitat fragmentation, and reduced hunting and birdwatching

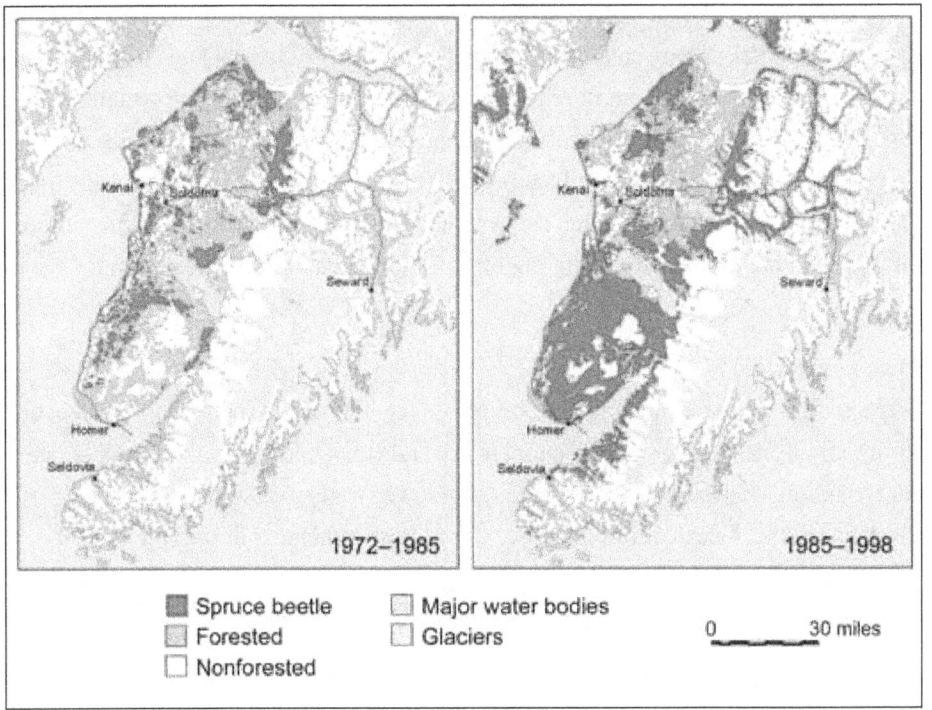

Figure 3-13—Spruce forest loss in Kenai Peninsula, Alaska. Source: USGCRP 2009 based on Berman et al. 1999.

could have adverse impacts on forest-sensitive rural communities in terms of lower employment and income.

Although current research suggests that timber supply will expand nationally, regional impacts are much more uncertain. A higher level of atmospheric CO_2 in the atmosphere results in trees capturing more carbon from the atmosphere and higher growth rates in some regions, especially in relatively young forests on fertile soils (Ryan et al. 2008). This increased growth could be tempered, however, by local conditions such as moisture stress, nutrient availability, or increased tropospheric ozone (Karnosky et al. 2005, Triggs et al. 2004). In the absence of dramatic increases in disturbance, effects of climate change could result in larger timber inventories (Perez-Garcia et al. 2002). Climate change scenarios predicting increased harvests, however, tend to lead to lower prices and, as a consequence, reduced harvests in regions with higher production costs (Perez-Garcia et al. 2002, Sohngen and Sedjo 2005). Warmer winters with more sporadic freezing and thawing are likely to increase erosion and landslides on forest roads, and reduce access for winter harvesting (USGCRP 2009), in turn, increasing cost and reducing supply of forest products. Under these conditions, a shrinking forest industry would lead to loss of employment for many rural communities.

The effect of climate change on forest-dependent communities will vary regionally. Wildfire risk is expected to be most severe in the Southwest and Northwest, largely because of higher summer temperatures and earlier spring snowmelt. Higher temperatures could lead to increased incidences of pest attacks and tree diseases all across the United States. Altered harvesting frequency and associated impacts on forest product prices could be felt nationwide. The impact, however, is expected to be higher in the timber-producing regions of the Southeast and old-growth forests of the West.

Water resources—

Impacts of climate change on water resources could result in increasing incidences of droughts, changing precipitation intensity and runoff, lower availability of water for irrigation, changing water demands, and lower water availability for energy production. Incidences of drought have dramatically changed during the last 50 years (fig. 3-14). Much of the Southeast and West have faced increases in drought severity and duration, while decreases have been observed in much of Midwest, Great Plains, and Northeast.

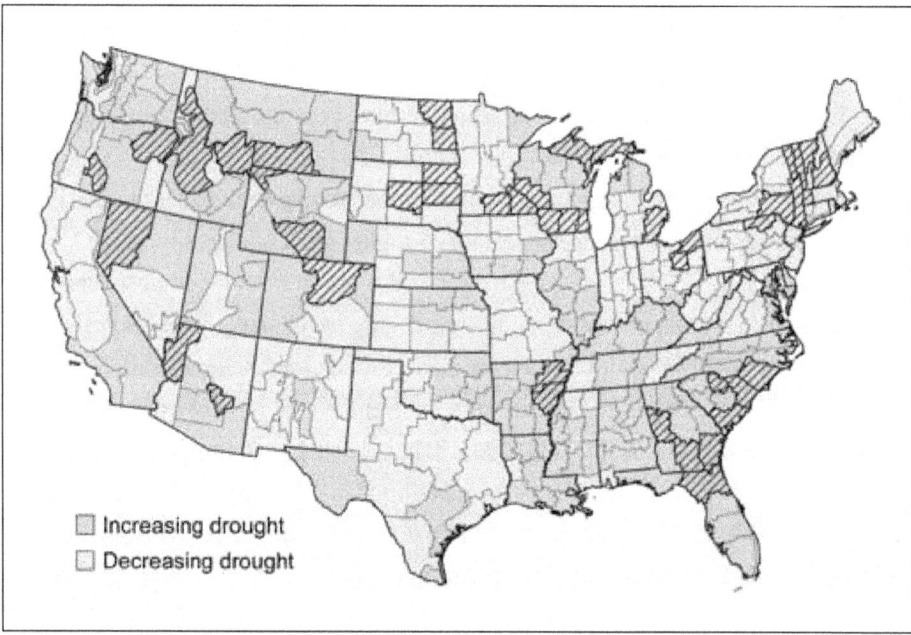

Figure 3-14— United States drought trends, 1958–2007. Crosshatched areas have statistically significant trends. Source: USGCRP 2009 based on Guttman and Quale 1996.

Limitations imposed on water supply by projected temperature increases are likely to be made worse by substantial reductions in rain and snowfall in the spring months, when precipitation is most needed to fill reservoirs to meet summer demand (Milly et al. 2008). The number of dry days between precipitation events is also projected to increase in the Southwest and the Mountain West, two of the most rapidly growing areas of the country. Continued population growth in these arid and semiarid regions would also stress water supplies, although the impact will be more severe for urban centers than rural counties.

Floods are also projected to be more frequent and intense as regional and seasonal precipitation patterns change and rainfall becomes more concentrated in heavy events. For the past century, total precipitation has increased by about 7 percent, while the heaviest 1 percent of rain events increased by approximately 20 percent (Gutowski et al. 2008). In general, International Panel for Climate Change climate models agree that northern areas are likely to get wetter and southern areas drier. Figure 3-15 outlines projected average precipitation changes by the 2090s in terms of light, moderate, and heavy storm events. The lightest precipitation is projected to decrease, and the heaviest will increase, continuing the observed trends.

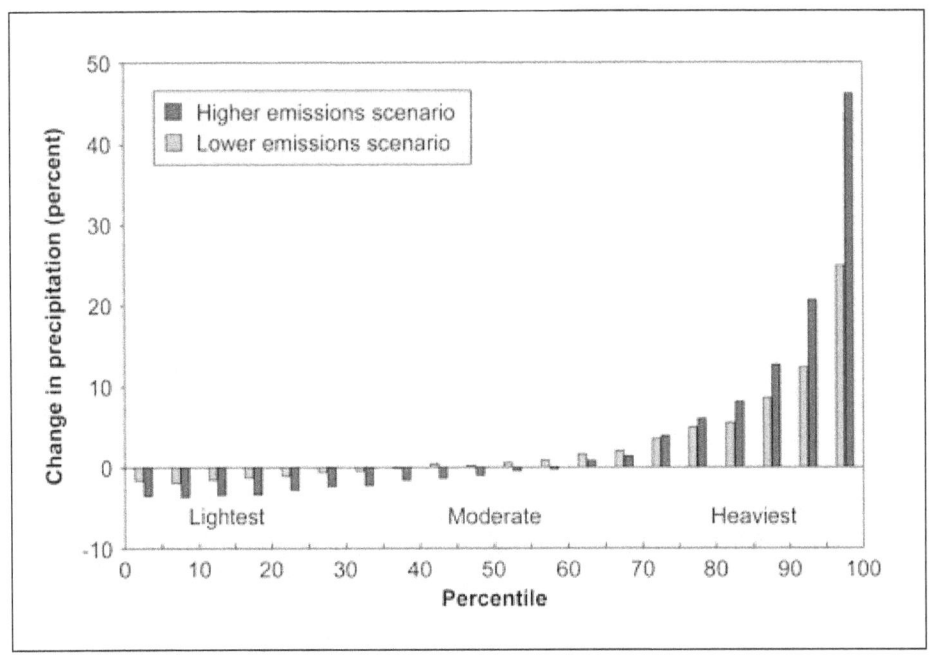

Figure 3-15—Projected changes in light, moderate, and heavy precipitation in North America (based on Intergovernmental Panel for Climate Change higher and lower emission scenarios). Source: USGCRP 2009 based on Gutowski et al. 2008.

Climate change is also projected to cause changes in runoff, the amount of precipitation that is not evaporated, stored as snowpack or soil moisture, or filtered down to groundwater. Figure 3-16 shows that the eastern part of the country will experience increased runoff, accompanied by declines in the West, especially the Southwest. This means that wet areas are projected to get wetter and dry areas drier, thus adding to the woes of agricultural and forest-dependent communities whose livelihoods (or incomes) in many cases are sensitive to water availability. The farming-dominated counties in the Great Plains and Midwest, however, are not expected to experience as large an impact as their Northeastern, Western, or Southwestern counterparts.

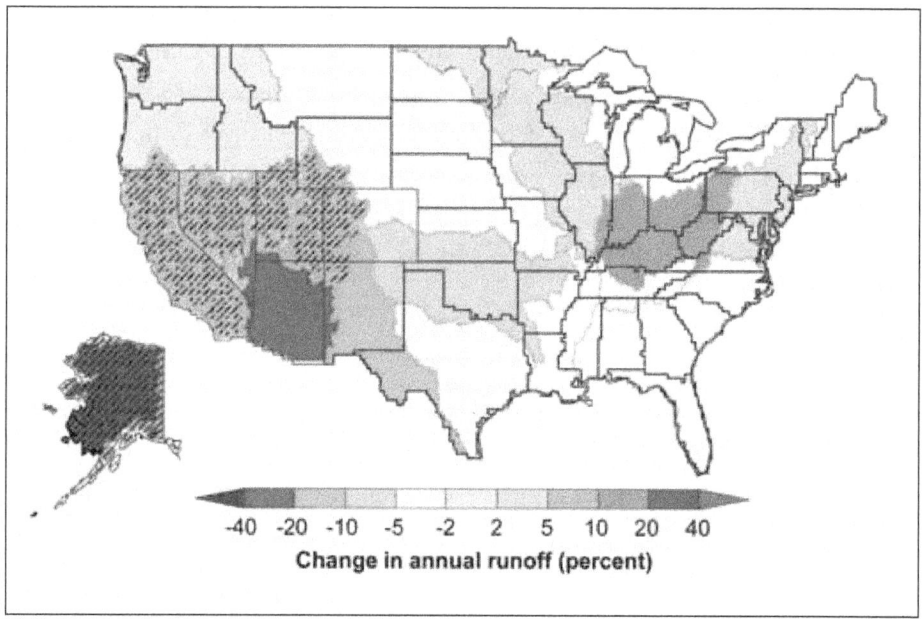

Figure 3-16—Projected changes in annual runoff for 2041–2060 relative to 1901–1970 for emissions in between the lower and higher emissions scenarios. Crosshatched areas indicate greater confidence in projection because of strong agreement among model projections. White areas indicate divergence among model projections. Source: USGCRP 2009:45 based on Milly et al. 2008.

Meeting the challenge of climate change has important implications for the United States in terms of intervention and resolution of intra- and intergroup conflicts. For example, decreased water availability in different regions of the country owing to increased temperature and lower or infrequent precipitation, along with an increase in water demand from increased population or agricultural activities, could produce more frequent and intense conflicts over water rights. The U.S. Bureau of Reclamation (2005) has identified many areas in the country that are already at risk for serious conflicts over water, even in the absence of climate change. Figure 3-17

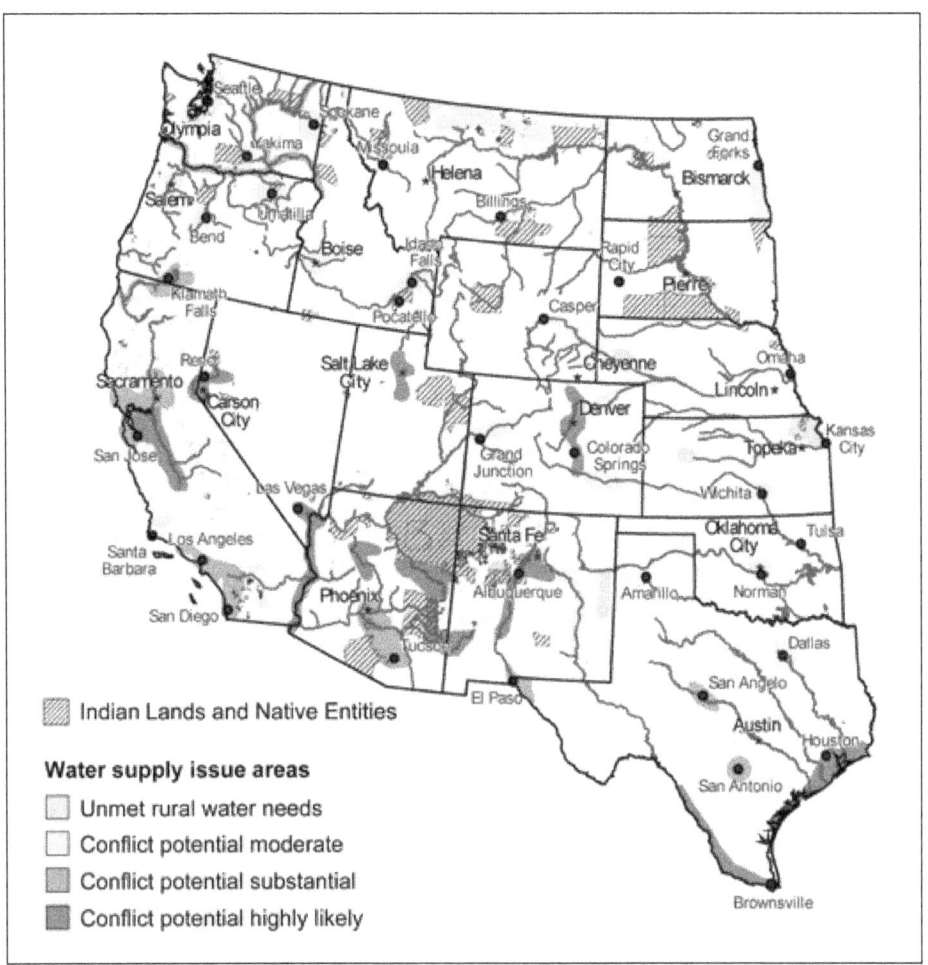

Figure 3-17—Likely water supply conflict regions in Western United States by 2025 without climate change effects. Source: USGCRP 2009 based on U.S. Bureau of Reclamation 2005.

shows regions in the West where water supply conflicts are likely to occur by 2025, based on a combination of factors such as population trends and endangered species need for water without factoring in climate change effects, which might exacerbate many of these conflicts.

Rural communities engaged in activities like farming are expected to be under additional water stress from climate change. For example, climate change increases the chance of water-related conflicts in already water-scarce regions like the Great Plains and Southwest. Current water use in the Great Plains is unsustainable, as the High Plains aquifer continues to be tapped faster than the rate of recharge. Similarly, groundwater pumping is lowering water tables, and rising temperatures reduce riverflows in vital rivers (Barnett et al. 2008)

Energy—

With increasing pressure on existing energy sources, rural communities' access to traditional energy sources could be threatened by climate change. Increased demand for energy as well as lower or uncertain supplies in many areas will accentuate such threats. Scott and Huang (2007) projected that temperature increases are likely to increase peak demand for electricity in most regions of the country. The energy demand for cooling is also expected to increase along with peak demands and higher temperatures. The increase in demand will be higher in the South, a region with especially high per capita electricity use (Scott and Huang 2007). The Southern region accounts for most of the persistently poor rural counties in the United States. Rural communities in the South might not be able to cope with the higher costs associated with increased demand, in turn increasing their vulnerability to climate change.

Renewable sources of energy, such as biomass-based energy, are already being promoted to increase energy supply, create jobs, reduce reliance on fossil fuels, and improve access to rural communities. The federal and state government incentives and mandates such as renewable fuel standards, blending incentives, research and development support, among others, are accelerating the process of making such energy sources commercially viable. The biomass-based energy markets could benefit rural landowners in terms of higher product prices as well as increased avenues for employment.

Fisheries—

America's coastlines and fisheries are especially at risk from climate change. Fisheries feed local people and provide livelihood to rural communities and indigenous peoples in many parts of the country. The habitats of some mountain species and coldwater fish, such as salmon and trout, are very likely to contract in response to warming, whereas some warm-water fishes such as smallmouth bass (*Micropterus dolomieu*) and bluegill (*Lepomis macrochirus*) might expand their ranges (Janetos et al. 2008). Apart from changes in species composition and availability of native fish species, aquatic ecosystem disruptions are likely to be compounded by invasion of nonnative invasive species, which tend to thrive under a wide range of environmental conditions.

In Alaska, climate change is already causing significant alterations in marine ecosystems (fig. 3-18), restricting fisheries and adding to the hardships of the rural people who depend on them (USGCRP 2009). Historically, warm periods in coastal waters have coincided with relatively low abundances of salmon, and cooler ocean temperatures have coincided with relatively high salmon numbers (Crozier et al. 2008). It has been estimated that as much as 40 percent of Northwest salmon populations may be lost by 2050 owing to climate change (Battin et al. 2007).

Climate change has caused alterations in marine ecosystems in Alaska.

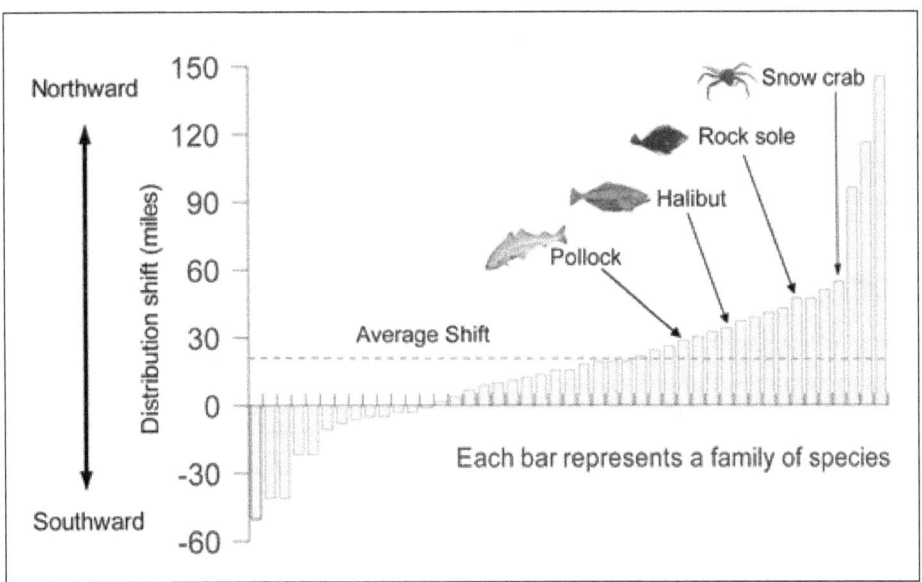

Figure 3-18—Marine species shift in Alaska from 1982 to 2006. Source: USGCRP 2009 based on Mueter and Litzow 2007.

Alaska leads the country in terms of its commercial and subsistence fishing catch. Most of the Nation's salmon, crab, halibut, and herring come from Alaska. In addition, many native communities depend on local harvests of fish, walruses, seals, whales, seabirds, and other marine species for their food supply. Subsistence fishing accounts for a large share of the food consumed in rural Alaska. The state's rural residents harvest an average of 225 pounds of fish per person (U.S. Fish and Wildlife Service 2010). The warmer water is already leading to a lower catch of salmon in Alaska creating hardships on the rural people and indigenous tribes who are dependent on these fishes for subsistence or employment (NTAA 2009).

In the Northeast, lobster fisheries are projected to continue a northward shift, and the cod fisheries in the Northeast are likely to be diminished with increasing ocean temperatures associated with climate change (USGCRP 2009). The possibility of ocean acidification owing to climate change may also endanger fisheries in the Northeast. For example, increased acidification in Passamaquoddy and Cobscook Bays and Bay of Fundy threaten shellfish including clams, scallops, and lobsters.

Climate change is also expected to reduce coral reefs and reef fish species (Graham et al. 2006). Changes in the species composition of coral reef ecosystems will likely have significant repercussions for both subsistence and commercial fishing in Hawaii, Puerto Rico, U.S. Virgin Islands, Guam, and American Samoa (Janetos et al. 2008). Change in fish availability owing to climate change may hit Guam and American Samoa particularly hard as almost all communities within the Pacific Islands derive between 25 and 69 percent of their animal protein from fish (Hotta 2000). The Southeast United States will most likely also observe a decline of wetland-dependent coastal fish and shellfish populations owing to the rapid loss of coastal marsh from rising sea levels associated with climate change (Zimmermann et al. 2002).

Conclusions

The potential impacts of climate change on rural communities include increased risks to human health, changes to the agricultural and forestry sectors, stress on water resources and fisheries, increased conflicts over scarce resources, impacts on recreation and tourism, adverse effects on indigenous communities, and additional impacts related to an increase in adverse weather events. Directly or indirectly, positively or negatively, climate change will affect all sectors and regions of the country, although the impacts will not be homogenous across regions, sectors, population groups, or time.

The impact of climate change on rural communities depends on complex interactions among different sectors, regions, and population groups and the environment. However, there is a dearth of information and literature on how the myriad socioeconomic and demographic factors will react to the biophysical changes accompanying climate change and virtually none on how the interconnected socioeconomic/ecological systems will respond. Most of the current literature is based on such coarse temporal and spatial resolution as to offer only very general guidance for investment and policymaking. For example, understanding the economic effects of climate change on timber production is constrained by limited scientific understanding of several key factors that control the response of natural and managed forests to climate change. Timber production will depend not only on climatic factors but also on stresses from pollution (e.g., acid rain), future trends in forest management practices, economic demand for forest products, and land-use change. Clarification of the uncertainties concerning how all of these factors will interact in the face of climate change will permit more informed policy and programmatic responses to reducing the vulnerability of rural communities to the impacts of climate change.

Climate change will affect rural communities through changes in availability or access to climate-sensitive resources that occur at local, regional, or national levels. The vitality of local communities (Hutton 2001, Jensen 2009, Wall et al. 2005), changes in monetary conditions (Ikeme 2003), status of emergency facilities and preparedness and planning (Murphy et al. 2005), condition of the public health system (Kinney et al. 2001), and exposure to conflict (Barnett 2003) all have the potential to either exacerbate or ameliorate the vulnerability of rural communities to climate change. Vulnerability to climate change tends to be greater for rural communities who typically have fewer resources and fewer alternatives than urban areas. Therefore, the climate risk mitigation and adaptive capacity of rural communities remains an important area for public policy interventions and future research. A suite of adaptation and mitigation policy options needs to be developed to reduce vulnerability of rural communities under a variety of climate change scenarios.

In light of the potential impacts of climate change on rural communities, enhancing their coping and adaptive capabilities is crucial. However, public discussion about adaptation is at an early stage in the United States (Moser 2005). An active dialogue among stakeholders and political institutions could help clarify the opportunities for adapting to and coping with climate change. A significant difference in infrastructure needs between rural and urban areas suggests that research focusing on assessing rural communities' adaptive capacity, costs and effectiveness of adaptation options, implementation impediments, and expected consequences is warranted.

Although much data on the biophysical impacts of climate change are already freely and readily available to a broad range of users, sociocultural and economic data and information related to how climate change will affect rural communities, their resilience, and adaptive capacity are scarce. Developing effective data collection systems and analyses of these issues first requires agreed-upon baseline indicators and measures of environmental, demographic, and economic conditions that can be used to track the effects of changes in climate on rural communities (Karl et al. 2009). A set of regional studies is needed to improve our understanding of climate change impacts and the distribution of costs and benefits of the impacts across rural and urban communities in the United States, and to develop appropriate policies to mitigate the impacts.

Acknowledgments

The authors thank the following reviewers for their helpful comments: Linda Langner, National Program Leader for the Resource Planning Act Assessment, J. Michael Bowker of the Southern Research Station, and Ellen M. Donoghue of Pacific Northwest Research Station, all of U.S. Forest Service, as well as G. Andrew Stainback of the Department of Forestry, University of Kentucky. Thanks also go to Jaganaddha Rao Matta, Forestry Officer of Finance, Food and Agricultural Organization of the United Nations for his helpful insights.

Metric Equivalents

When you know:	Multiply by:	To get:
Acres (ac)	0.405	Hectares
Pounds (lb)	454	Grams
Tons	.907	Tonnes (metric tons)
Degrees Farenheit (°F)	.55(°F − 32)	Degrees Celsius

Literature Cited

Alig, R.J.; Adams, D.; Joyce, L.; Sohngen, B. 2004. Climate change impacts and adaptation in forestry: responses by trees and markets. Choices. Fall: 7–11.

Barnett, J. 2003. Security and climate change. Global Environmental Change. 13: 7–17.

Barnett, T.P.; Pierce, D.W.; Hidalgo, H.G. [et al.]. 2008. Human-induced changes in the hydrology of the Western United States. Science. 319(5866): 1080–1083.

Battin, J.; Wiley, M.W.; Ruckelshaus, M.H. [et al.]. 2007. Projected impacts of climate change on salmon habitat restoration. Proceedings of the National Academy of Sciences. 104(16): 6720–6725.

Berman, M.; Juday, G.P.; Burnside, R. 1999. Climate change and Alaska's forests: people, problems, and policies. In: Weller, G.; Anderson, P.A., eds. Assessing the consequences of climate change in Alaska and the Bering Sea Region. Proceedings of a workshop. Fairbanks, AK: Center for Global Change and Arctic System Research, University of Alaska: 21–42.

Bindoff, N.L.; Willebrand, J.; Artale, V. [et al.]. 2007. Observations: oceanic climate change and sea level. In: Solomon, S.; Qin, D.; Manning, M. [et al.], eds. Climate change 2007: the physical science basis. Contribution of working group I to the fourth assessment report of the Intergovernmental Panel on Climate Change. Cambridge, United Kingdom, and New York: Cambridge University Press: 385–432.

Brown, T.J.; Hall, B.L.; Westerling, A.L. 2004. The impact of twenty-first century climate change on wildland fire danger in the Western United States: an applications perspective. Climatic Change. 62: 365–388.

Burkett, V.; Ritschard, R.; McNulty, S. [et al.]. 2001. Potential consequences of climate variability and change for the Southeastern United States. Report for the U.S. Global Change Research Program. Cambridge, United Kingdom: Cambridge University Press: 137–164.

Cameron, G.N.; Scheel, D. 2001. Getting warmer: effect of global climate change on distribution of rodents in Texas. Journal of Mammalogy. 82(3): 652–680.

Christensen, J.H.; Hewitson, B.; Busuioc, A. [et al.]. 2007. Regional climate projections. In: Solomon, S.; Qin, D.; Manning, M. [et al.], eds. Climate change 2007: the physical science basis. Contribution of working group I to the fourth assessment report of the Intergovernmental Panel on Climate Change. Cambridge, United Kingdom, and New York: Cambridge University Press: 847–940.

Clark, P.U.; Weaver, A.J.; Brook, E. [et al.]. 2008. Abrupt climate change. Synthesis and Assessment Product 3.4. Reston, VA: U.S. Geological Survey. 459 p.

Cromartie, J.; Bucholtz, S. 2008. Defining the "rural" in rural America. Amber Waves. June 2008. http://www.ers.usda.gov/AmberWaves/June08/ Features/RuralAmerica.htm. (25 May 2010).

Crozier, L.G.; Hendry, A.P.; Lawson, P.W. [et al.]. 2008. Potential responses to climate change in organisms with complex life histories: evolution and plasticity in Pacific salmon. Evolutionary Applications. 1(2): 252–270.

Deller, S.C.; Tsung-Hsiu T.; Marcouiller, D.W. [et al.]. 2001. The role of amenities and quality of life in rural economic growth. American Journal of Agricultural Economics. 83(2): 352–365.

Donoghue, E.; Lynn, K.; MacKendrick, K.; Belanger, K. 2009. Climate change and social vulnerability: a briefing paper for policymakers and land managers. Draft October 2009. On file with: E.M. Donoghue, research social scientist, Forestry Sciences Laboratory, 620 SW Main St., Suite 400, Portland, OR 97205.

Easterling, D.R.; Karl, T.R. 2001. Potential consequences of climate variability and change for the Midwestern United States. In: Climate change impacts on the United States: the potential consequences of climate variability and change. Report for the U.S. Global Change Research Program. Cambridge, United Kingdom: Cambridge University Press: 167–188. Chapter 6.

Easterling, W.E.; Aggarwal, P.K.; Batima, P. [et al.]. 2007. Food, fibre and forest products. In: Parry, M.L.; Canziani, O.F.; Palutikof, J.P. [et al.], eds. Climate change 2007: impacts, adaptation and vulnerability. Contribution of working group II to the fourth assessment report of the Intergovernmental Panel on Climate Change. Cambridge, United Kingdom: Cambridge University Press: 273–313.

Ebi, K.L.; Balbus, J.; Kinney, P.L. 2008. Effects of global change on human health. In: Gamble, J.L., ed. Analyses of the effects of global change on human health and welfare and human systems. Synthesis and Assessment Product 4.6. Washington, DC: U.S. Environmental Protection Agency: 39–87.

Elliott, G.P.; Baker, W.L. 2004. Quaking aspen (*Populus tremuloides* Michx.) at treeline: a century of change in the San Juan Mountains, Colorado, USA. Journal of Biogeography. 31: 733–745.

Feng, S.; Hu, Q. 2007. Changes in winter snowfall/precipitation ratio in the contiguous United States. Journal of Geophysical Research. 112: D15109. DOI: 10.1029/2007JD008397.

Field, C.B.; Mortsch, L.D.; Brklacich, M. [et al.]. 2007. North America. In: Parry, M.L.; Canziani, O.F.; Palutikof, J.P. [et al.], eds. Climate change 2007. Impacts, adaptation and vulnerability. Contribution of working group II to the fourth assessment report of the Intergovernmental Panel on Climate Change. Cambridge, United Kingdom, and New York: Cambridge University Press: 617–652.

Flint, C.G.; Luloff, A.E. 2005. Natural resource based communities, risk and disaster: an intersection of theories. Society and Natural Resources.18: 399–412.

Frumhoff, P.C.; McCarthy, J.J.; Melillo, J.M. [et al.]. 2007. Confronting climate change in the U.S. Northeast: science, impacts and solutions. Synthesis report of the Northeast Climate Impacts Assessment. Cambridge, MA: Union of Concerned Scientists. 146 p.

Gan, J. 2004. Risk and damage of southern pine beetle outbreaks under global climate change. Forest Ecology and Management. 191(1-3): 61–71.

Graham, N.A.J.; Wilson, S.K.; Jennings, S. [et al.]. 2006. Dynamic fragility of oceanic coral reef ecosystems. Proceedings of the National Academy of Sciences. 103(22): 8425–8429.

Gu, L.; Hanson, P.J.; Post, W.M. 2008. The 2007 Eastern U.S. spring freeze: Increased cold damage in a warming world? BioScience. 58(3): 253–262.

Gutowski, W.J.; Hegerl, G.C.; Holland, J. 2008. Causes of observed changes in extremes and projections of future changes. In: Karl, T.R.; Meehl, G.A.; Miller, C.D. [et al.], eds. Weather and climate extremes in a changing climate: regions of focus: North America, Hawaii, Caribbean, and U.S. Pacific Islands. Synthesis and Assessment Product 3.3. Washington, DC: U.S. Climate Change Science Program: 81–116.

Guttman, N.B.; Quayle, R.G. 1996. A historical perspective of U.S. climate divisions. Bulletin of the American Meteorological Society. 77(2): 293–303.

Hanna, J.M. 2007. Native communities and climate change: legal and policy approaches to protect tribal legal rights. Boulder, CO: Natural Resource Law Center, University of Colorado School of Law. 66 p.

Hatfield, J.; Boote, K.; Fay, P. [et al.]. 2008. Agriculture. In: Backlund, P.; Janetos, A.; Schimel, D. [et al.], eds. The effects of climate change on agriculture, land resources, water resources, and biodiversity in the United States. Synthesis and Assessment Product 4.3. Washington, DC: U.S. Climate Change Science Program: 21–74.

Hayhoe, K.; Cayan, D.; Field, C. 2004. Emissions pathways, climate change, and impacts on California. Proceedings of the National Academy of Sciences. 101: 12422–12427.

Heileman, S.; Walling, L.J.; Douglas, C. [et al.], eds. 2004. Caribbean environment outlook. Kingston, Jamaica: United Nations Environmental Programme. 114 p. http://www.unep.org/geo/pdfs/caribbean_eo.pdf. (7 June 2010).

Hotta, M. 2000. The sustainable contribution of fisheries to food security in the Asia and Pacific region: regional synthesis. In: Sustainable contribution of fisheries to food security. Bangkok, Thailand: Food and Agriculture Organization of the United Nations: 1–28.

Houser, S.; Teller, V.; MacCracken, M. [et al.]. 2001. Potential consequences of climate variability and change for native peoples and homelands. In: Climate change impacts on the United States: the potential consequences of climate variability and change. Report for the U.S. Global Change Research Program. U.S. National Assessment Synthesis Team. Cambridge, United Kingdom: Cambridge University Press: 351–377.

Hutton, D. 2001. Psychosocial aspects of disaster recovery: integrating communities into disaster planning and policy making. Toronto, ON: Institute for Catastrophic Loss Reduction. 16 p.

Ikeme, J. 2003. Equity, environmental justice and sustainability: incomplete approaches in climate change politics. Global Environmental Change. 13: 195–206.

Intergovernmental Panel for Climate Change [IPCC]. 2001. Climate change: impacts, adaptation, and vulnerability. In: McCarthy, J.J.; Canziani, O.F.; Leary, N.A. [et al.], eds. Contribution of working group II to the third assessment report of the Intergovernmental Panel on Climate Change. Cambridge, United Kingdom and New York: Cambridge University Press: 1–103.

Intergovernmental Panel for Climate Change [IPCC]. 2007. Climate change: impacts, adaptation, and vulnerability. In: Parry, M.L.; Canziani, O.F.; Palutikof, J.P. [et al.], eds. Contribution of working group II to the fourth assessment report of the Intergovernmental Panel on Climate Change. Cambridge, United Kingdom: Cambridge University Press: 1–131.

Janetos, A.; Hansen, L.; Inouye, D. [et al.]. 2008. Biodiversity. In: Backlund, P.; Janetos, A.; Schimel, D., eds. The effects of climate change on agriculture, land resources, water resources, and biodiversity in the United States. Synthesis and Assessment Product 4.3. U.S. Washington, DC: Department of Agriculture: 151–181.

Jensen, K.J. 2009. Climate change and rural communities in the U.S. Draft briefing paper. Rural Policy Research Institute. http://www.rupri.org/Forms/Climate_Change_Brief.pdf. (11 June 2010).

Johnson, K.M.; Beale, C.L. 2002. Nonmetro recreation counties: their identification and rapid growth. Rural America. 17(4): 12–19.

Johnstone, J.; Chapin, F.S., III. 2003. Non-equilibrium succession dynamics indicate continued northern migration of lodgepole pine. Global Change Biology. 9: 1401–1409.

Jones, C.A.; Parker, T.S.; Ahearn, M. [et al.]. 2009. Health status and health care access of farm and rural populations. Economic Info. Bull. 57. Washington, DC: Department of Agriculture. 72 p.

Karl, T.R.; Melillo, J.M.; Peterson, T.C., eds. 2009. Global climate change impacts in the United States. Cambridge, United Kingdom: Cambridge University Press. 196 p.

Karnosky, D.F.; Pregitzer, K.S.; Zak, D.R. [et al.]. 2005. Scaling ozone responses of forest trees to the ecosystem level in a changing climate. Plant Cell Environment. 28: 965–981.

Kiely, T.; Donaldson, D.; Grube, A. 2004. Pesticides industry sales and usage: 2000 and 2001 market estimates. Washington, DC: U.S. Environmental Protection Agency. 33 p.

Kilpatrick, A.M.; Meola, M.A.; Moudy, R.M.; Kramer, L.D. 2008. Temperature, viral genetics, and the transmission of West Nile virus by *Culex pipiens* mosquitoes. PLoS Pathogens. 4(6): e1000092. DOI: 10.1371/journal. ppat.1000092. (11 June 2010).

Kinney, P.L.; Shindell, D.; Chae, E.; Winston, B. 2001. Public health. In: Rosenzweig, R.; Solecki, W.D., eds. Climate change and a global city: the potential consequences of climate variability and change. New York: Columbia Earth Institute: 103–120.

Kleinosky, L.R.; O'Sullivan, D.; Yarnal, B. 2005. A method for constructing a social vulnerability index: an application to hurricane storm surges in a developed country. Mitigation Adaptation Strategies for Global Change. 11: 741–764.

Kunkel, K.E.; Huang, H.-C.; Liang, X.-Z. [et al.]. 2008. Sensitivity of future ozone concentrations in the Northeast U.S. to regional climate change. Mitigation and Adaptation Strategies for Global Change. 13(5-6): 597–606.

Lazar, B.; Williams, M. 2008. Climate change in western ski areas: potential changes in the timing of wet avalanches and snow quality for the Aspen ski areas in the years 2030 and 2100. Cold Regions Science and Technology. 51(2-3): 219–228.

Levia, D.F.; Frost, E.E. 2004. Assessment of climatic suitability for the expansion of *Solenopsis invicta* Buren in Oklahoma using three general circulation models. Theoretical and Applied Climatology. 79(1-2): 23–30.

Lobell, D.; Field, C.; Cahill, K.N.; Bonfils, C. 2006. Impacts of future climate change on California perennial crop yields: model projections with climate and crop uncertainties. Agricultural and Forest Meteorology. 141(2-4): 208–218.

Logan, J.A.; Regniere, J.; Powell, J.A. 2003. Assessing the impacts of global warming on forest pest dynamics. Frontiers in Ecology and Environment. 1: 130–137.

Milly, P.C.D.; Betancourt, J.; Falkenmark, M. [et al.]. 2008. Stationarity is dead: Whither water management? Science. 319(5863): 573–574.

Mimura, N.; Nurse, L.; McLean, R.F. [et al.]. 2007. Small islands. In: Parry, M.L.; Canziani, O.F.; Palutikof, J.P. [et al.], eds. Climate change 2007. Impacts, adaptation and vulnerability. Contribution of working group II to the fourth assessment report of the Intergovernmental Panel on Climate Change. Cambridge, United Kingdom: Cambridge University Press: 687–716.

Morello F.; Rachel, M.P; Sadd, J.; Shonkoff, S. 2009. The climate gap: inequalities in how climate change hurts Americans and how to close the gap. http://college.usc.edu/pere/documents/The_Climate_Gap_Full_Report_FINAL. pdf. (12 June 2010).

Moser, S. 2005. Enhancing decision-making through integrated climate research. Summary of an exploratory workshop for the NOAA-OGP-RISA Program, Alaska regional meeting. Anchorage, AK: National Oceanic and Atmospheric Administration—Office of Global Programs. 63 p.

Motha, R.P.; Baier, W. 2005. Impacts of present and future climate change and climate variability on agriculture in the temperate regions: North America. Climatic Change. 70(1-2): 137–164.

Mueter, F.J.; Litzow, M.A. 2007. Sea ice retreat alters the biogeography of the Bering Sea continental shelf. Ecological Applications. 18(2): 309–320.

Murphy, B.; McBean, G.; Dolan, H. [et al.]. 2005. Enhancing local level emergency management: the influence of disaster experience and the role of household and neighbourhoods. ICLR Res. Pap. 43. Toronto, Canada: Institute for Catastrophic Loss Reduction. 79 p.

National Assessment Synthesis Team [NAST]. 2001. Climate change impacts on the United States: the potential consequences of climate variability and change. Report for the U.S. Global Change Research Program. Cambridge, United Kingdom: Cambridge University Press. 620 p.

National Tribal Air Association [NTAA]. 2009. Impacts of climate change on tribes in the United States. Submitted December 11, 2009 to Assistant Administrator Gina McCarthy, USEPA, Office of Air and Radiation. http://www.epa.gov/air/tribal/pdfs/Impacts%20of%20Climate%20Change%20on%20Tribes%20in%20the%20United%20States.pdf. (12 June 2010).

Nilsson, C. 2008. Climate change from an indigenous perspective: key issues and challenges. Indigenous Affairs. 1–2: 9–16.

Parson, E.A.; Carter, L.; Anderson, P. [et al.]. 2001. Potential consequences of climate variability and change for Alaska. In: National Assessment Synthesis Team, ed. Climate change impacts on the United States. Cambridge, United Kingdom: Cambridge University Press: 283–312.

Parton, W.; Gutmann, M.; Ojima, D. 2007. Long-term trends in population, farm income, and crop production in the Great Plains. BioScience. 57(9): 737–747.

Perez-Garcia, J.; Joyce, L.A.; McGuire, A.D.; Xiao, X. 2002. Impacts of climate change on the global forest sector. Climatic Change. 54: 439–461.

Peterson, A.T. 2003. Projected climate change effects on Rocky Mountain and Great Plains birds: generalities of biodiversity consequences. Global Change Biology. 9(5): 647–655.

Physicians for Social Responsibility [PSR]. 2010. Health implications of global warming: vector-borne and water-borne diseases. http://www.psr.org/assets/pdfs/vector-borne-and-water-borne.pdf. (13 June 2010).

Rahmstorf, S.; Cazenave, A.; Church, J.A. [et al.]. 2007. Recent climate observations compared to projections. Science. 316(5825): 709.

Reeder, R.J.; Brown, D.M. 2005. Recreation, tourism, and rural well-being. Economic Res. Rep. 7. Washington, DC: U.S. Department of Agriculture. 33 p.

Richardson, R.B.; Loomis, J.B. 2004. Adaptive recreation planning and climate change: a contingent visitation approach. Ecological Economics. 50: 83–99.

Rosenzweig, C.; Solecki, W., eds. 2001. Climate change and a global city: the potential consequences of climate variability and change—Metro East Coast. New York: Columbia Earth Institute. 224 p.

Ruth, M.; Coelho, D.; Karetnikov, D. 2007. The U.S. economic impacts of climate change and the costs of inaction. College Park, MD: Center for Integrative Environmental Research, University of Maryland. 52 p.

Ryan, M.G.; Archer, S.R.; Birdsey, R. [et al.]. 2008. Land resources. In: Backlund, P.; Janetos, A.; Schimel, D. [et al.], eds. The effects of climate change on agriculture, land resources, water resources, and biodiversity in the United States. Synthesis and Assessment Product 4.3. Washington, DC: U.S. Department of Agriculture: 75–120.

Scott, D.; Dawson, J.; Jones, B. 2008. Climate change vulnerability of the U.S. Northeast winter recreation–tourism sector. Mitigation and Adaptation Strategies for Global Change. 13(5-6): 577–596.

Scott, D.; Jones, B. 2005. Climate change and Banff National Park: implications for tourism and recreation Waterloo, Ontario. 31 p. Report prepared for the Town of Banff. Waterloo, ON: University of Waterloo. http://www. fes.uwaterloo.ca/ geography/faculty/danielscott/PDFFiles/BANFF_Final%20copy_MAY%202006. pdf. (15 June 2010).

Scott, D.; Jones, B. 2006. Climate change and nature-based tourism: implications for park visitation in Canada. Waterloo, ON: University of Waterloo, Department of Geography. http://www.fes.uwaterloo.ca/geography/faculty/danielscott/ PDFFiles/NATURE_Final%20copy.pdf. (15 June 2010).

Scott, D.; Jones, B.; Konopek, J. 2007. Implications of climate and environmental change for nature-based tourism in the Canadian Rocky Mountains: a case study of Waterton Lakes National Park. Tourism Management. 28: 570–579.

Scott, M.J.; Huang, Y.J. 2007. Effects of climate change on energy use in the United States. In: Wilbanks, T.J.; Bhatt, V.; Bilello, D. E. [et al.], eds. Effects of climate change on energy production and use in the United States. Synthesis and Assessment Product 4.5. Washington, DC: U.S. Climate Change Science Program: 8–44.

Smit, B.; Pilifosova, O.; Burton, I. [et al.]. 2001. Adaptation to climate change in the context of sustainable development and equity. In: McCarthy, J.J.; Canziani, O.; Leary, N.A. [et al.], eds. Contribution of the working group II to the third assessment report of the Intergovernmental Panel on Climate Change. Cambridge, United Kingdom: Cambridge University Press: 877–912.

Sohngen, B.; Sedjo, R. 2005. Impacts of climate change on forest product markets: implications for North American producers. Forest Chronicles. 81: 669–674.

Sussman, F.G.; Cropper, M.L.; Galbraith, H. [et al.]. 2008. Effects of global change on human welfare. In: Gamble, J.L., ed. Analyses of the effects of global change on human health and welfare and human systems. Synthesis and Assessment Product 4.6. Washington, DC: U.S. Environmental Protection Agency: 111–168.

Triggs, J.M.; Kimball, B.A.; Pinter, P.A., Jr. [et al.]. 2004. Free-air CO_2 enrichment effects on the energy balance and evapotranspiration of sorghum. Agricultural and Forest Meteorology. 124(2): 63–79.

Tsosie, R. 2007. Indigenous people and environmental justice: the impact of climate change. University of Colorado Law Review. 78(4): 1625–1678.

Turner, B.L., II; Kasperson, R.E.; Matson P.A. [et al.]. 2003. A framework for vulnerability analysis in sustainability science. Proceedings of the National Academy of Science. 100: 8074–8079.

U.S. Bureau of Reclamation [USBR]. 2005. Water 2025: preventing crises and conflict in the West. Washington, DC. 32 p.

U.S. Department of Agriculture, Economic Research Service [USDA ERS]. 2010a. An enhanced quality of life for rural Americans: rural gallery. Nonmetro recreation counties, 1999. http://www.ers.usda.gov/Emphases/rural/gallery/NonmetroRecreation.htm. (11 June 2010).

U.S. Department of Agriculture, Economic Research Service [USDA ERS]. 2010b. An enhanced quality of life for rural Americans: rural gallery. Poverty rates by region and metro status, 2006. http://www.ers.usda.gov/emphases/rural/gallery/IncomePoverty.htm. (10 June 2010).

U.S. Department of Agriculture, Economic Research Service [USDA ERS]. 2010c. Farming dependent counties 1998–2000. http://www.ers.usda.gov/briefing/rurality/typology/maps/Farming.htm. (12 June 2010).

**U.S. Department of Agriculture, Economic Research Service [USDA ERS].
2010d.** Measuring rurality: 2004 county typology codes. http://www.ers.usda.
gov/Briefing/Rurality/Typology/. (13 June 2010).

**U.S. Department of Agriculture, Economic Research Service [USDA ERS].
2010e.** Measuring rurality: 2004 county typology codes methods, data sources,
and documentation. http://www.ers.usda.gov/Briefing/Rurality/Typology/
Methods/. (12 June 2010).

**U.S. Department of Agriculture, Economic Research Service [USDA ERS].
2010f.** Measuring rurality: What is rural? http://www.ers.usda.gov/Briefing/
Rurality/WhatIsRural/. (12 June 2010).

**U.S. Department of Agriculture, Economic Research Service [USDA ERS].
2010g.** Rural definitions. http://www.ers.usda.gov/Data/RuralDefinitions/. (15
June 2010).

**U.S. Department of Agriculture, Economic Research Service [USDA ERS].
2010h.** Rural income, poverty and welfare: nonfarm earnings. http://www.ers.
usda.gov/briefing/incomepovertywelfare/nonfarmearnings/. (11 June 2010).

**U.S. Department of Agriculture, Economic Research Service [USDA ERS].
2010i.** Rural income, poverty and welfare: poverty geography. http://www.ers.
usda.gov/briefing/incomepovertywelfare/PovertyGeography/. (6 October 2010).

**U.S. Department of Agriculture, Economic Research Service [USDA ERS].
2010j.** Rural income, poverty and welfare: rural income. http://www.ers.usda.
gov/Briefing/incomepovertywelfare/RuralIncome/. (12 June 2010).

**U.S. Department of Agriculture, Economic Research Service [USDA ERS].
2010k.** Rural income, poverty and welfare: rural welfare. http://www.ers.usda.
gov/Briefing/incomepovertywelfare/RuralWelfare/. (10 June 2010).

**U.S. Department of Agriculture, Economic Research Service [USDA
ERS]. 2010l.** Rural labor and education. http://www.ers.usda.gov/Briefing/
LaborAndEducation/. (15 June 2010).

U.S. Department of Agriculture, Forest Service [USFS]. 2000. Quantitative
environmental learning project. Data set #060 USFS emergency fire suppression.
Seattle, WA: Seattle Central Community College. http://www.seattlecentral.org/
qelp/sets/060/060. (13 June 2010).

U.S. Environmental Protection Agency [USEPA]. 2009. Frequently asked questions about global warming and climate change: back to basics. Office of Air and Radiation, April 2009. http://www.epa.gov/climatechange/downloads/Climate_Basics.pdf. (13 June 2010).

U.S. Fish and Wildlife Service. 2010. Federal subsistence management program. http://alaska.fws.gov/asm/about.cfml. (15 June 2010).

U.S. General Accounting Office [USGAO]. 2003. Alaska Native villages: most are affected by flooding and erosion, but few qualify for federal assistance. GAO-04-142. Washington, DC. 82 p. http://purl.access.gpo.gov/GPO/LPS42077. (11 June 2010).

U.S. Geological Survey [USGS]. 2005. Changes in streamflow timing in the Western United States in recent decades. USGS fact sheet 2005-3018. La Jolla, CA: U.S. Geological Survey, National Streamflow Information Program. 4 p. http://pubs.usgs.gov/fs/2005/3018/. (10 June 2010).

U.S. Global Change Research Program [USGCRP]. 2009. Global climate change impacts in the United States. Cambridge, United Kingdom: Cambridge University Press. 196 p.

Vermeer, M.; Rahmstorf, S. 2009. Global sea level linked to global temperature. Proceedings of the National Academy of Sciences. 106(51): 21527–21532.

Wall, E.; Smit, B.; Wandell, J. 2005. From silos to synthesis: interdisciplinary issues for climate change impacts and adaptation research. Canadian Association of Geographers special session series: communities and climate change impacts, adaptation and vulnerability, agriculture. Moncton, New Brunswick: Canadian-Climate Impacts and Adaptation Research Network. 24 p.

Westerling A.L.; Hidalgo, H.G.; Cayan, D.R.; Swetnam, T.W. 2006. Warming and earlier spring increase Western U.S. forest wildfire activity. Science. 313(5789): 940–943.

Wilbanks, T.J.; Lankao, P.R.; Bao, M. [et al.]. 2007. Industry, settlement and society. In: Parry, M.L.; Canziani, O.F.; Palutikof, J.P. [et al.], eds. Climate change 2007: impacts, adaptation and vulnerability. Contribution of working group II to the fourth assessment report of the Intergovernmental Panel on Climate Change. Cambridge, United Kingdom, and New York: Cambridge University Press: 357–390.

Williams, J.E.; Haak, A.L.; Gillespie, N.G. 2007. Healing troubled waters: preparing trout and salmon habitat for a changing climate. Arlington, VA: Trout Unlimited. 12 p. http://www.tu.org/climatechange. (13 June 2010).

Williams, S.J.; Gutierrez, B.T.; Titus, J.G. [et al.]. 2009. Sea-level rise and its effects on the coast. In: Titus, J.G.; Anderson, K.E.; Cahoon, D.R. [et al.], eds. Synthesis and Assessment Product 4.1. Washington, DC: U.S. Environmental Protection Agency: 11–24.

Whitener, L.A.; Parker, T. 2007. Policy options for a changing rural America. Amber Waves. 5(Spec. issue): 58–65. Originally published in 2005 3(2): 28–35.

Wolfe, W.; Ziska, L.; Petzoldt, C. [et al.]. 2007. Projected change in climate thresholds in the Northeastern U.S.: implications for crops, pests, livestock, and farmers. Mitigation and Adaptation Strategies for Global Change. 13(5-6): 555–575.

Wu, S.Y.; Yarnal, B.; Fisher, A. 2002. Vulnerability of coastal communities to sea-level rise: a case study of Cape May County, New Jersey, USA. Climate Resources. 22: 255–270.

Yohe, G.W.; Tol, R.S.J. 2002. Indicators for social and economic coping capacity—moving towards a working definition of adaptive capacity. Global Environmental Change. 12(1): 25–40.

Zimmerman, G.; O'Brady, C.; Hurlbutt, B. 2006. Climate change: modeling a warmer Rockies and assessing the implications. The 2006 state of the Rockies report card. Colorado Springs, CO: Colorado College: 89–102.

Zimmerman, R.J.; Minello, T.J.; Rozas, L.P. 2002. Salt marsh linkages to productivity of penaeid shrimps and blue crabs in the northern Gulf of Mexico. In: Weinstein, M.P.; Kreeger, D.A., eds. Concepts and controversies in tidal marsh ecology. Kluwer, Dordrecht and Boston, MA: Kluwer Academic Publishers: 293–314.

Ziska, L.H. 2003. Evaluation of the growth response of six invasive species to past, present, and future atmospheric carbon dioxide. Journal of Experimental Botany. 54(381): 395–404.

Chapter 4: Competiveness of Carbon Offset Projects on Nonindustrial Private Forest Lands in the United States

D. Evan Mercer, Pankaj Lal, and Janaki Alavalapati

Introduction

Growing concerns over the impacts of human-induced greenhouse gas (GHG) emissions have led U.S. policymakers to consider creating markets to regulate GHG emissions, although no bill has yet become law. For example, the House passed the American Clean Energy and Security Act (aka Waxman-Markey) on June 26, 2009, and three bills were submitted to the Senate in 2009 and 2010: the Clean Energy Jobs and American Power Act (Kerry-Boxer), the American Power Act (Kerry-Lieberman), and the Carbon Limits and Energy for America's Renewal Act (Cantwell-Collins). Waxman-Markey, Kerry-Boxer, and Kerry-Lieberman would create markets for emitting and offsetting carbon dioxide (CO_2) and permit the purchase of up to 2 billion metric tons of carbon offsets annually.[1] Cantwell-Collins would not allow emitters to purchase offsets; however, it would provide for the establishment of a trust fund to provide incentives—loans and grants to fund offset-like projects that reduce, avoid, or sequester GHG emissions through forestry and other land use initiatives. Appendix I provides a list of the eligible offset projects allowed in the 2010 American Powers Act (Kerry-Lieberman), the only current legislation to include a list of eligible offset projects. In addition to forestry and agricultural offset projects, potential offset projects include carbon capture and storage, methane collection, recycling and waste minimization, biochar production, and a variety of land management changes.

Parry et al. (2007) estimated that 10 to 20 percent of the world's anticipated GHG emissions could be offset over the next 50 years through forest preservation, tree planting, and improved farming methods. Currently, forests, urban trees, and agricultural soils offset approximately 15 percent of total U.S. CO_2 emissions from the energy, transportation, and other sectors (USEPA 2009a). McCarl and Schneider (2001) suggested that an additional 3 to 5 percent of carbon sequestration per year could be achieved through changes in agricultural and forest management, tree planting, and biofuel substitution. However, the market for U.S.-based forest offset projects in the current, voluntary over-the-counter market remains small. For

---

[1] Waxman-Markey allows 1 billion metric tons of offsets from domestic sources and 1 billion from international sources, and Kerry-Boxer allows 1.5 billion metric tons from domestic and 0.5 billion metric tons from international sources. If domestic sources are unable to satisfy the demand for offsets, however, the cap on international offsets can be raised to 1.5 billion metric tons under Waxman-Markey and 1.25 billion under the Kerry-Boxer bill.

example, as shown in table 4-1, U.S.-based forestry carbon projects account for only 4 percent of all carbon offset projects worldwide, 17 percent of all U.S.-based offset projects, and 15 percent of forestry projects worldwide (Carbon Catalog 2010).

Table 4-1—Number of carbon offset projects by type: total worldwide, total in United States, and percentage United States represents for each category

Offset project type	Worldwide Number of projects	United States Number of projects	U.S. Percentage of category
			Percent
Wind power	125	27	22
Forestry	114	17	15
Methane from animal biomass	67	22	33
Industrial methane	36	13	36
Fuel efficiency	25	2	8
Solar power	19	6	32
Fuel substitution	18	3	17
Efficient lighting	9	1	11
Efficient buildings	8	1	13
Hydroelectric power	6	1	17
Heat-electricity cogeneration	4	1	25
Material substitution	4	2	50
Public transportation	3	2	67
Geothermal	1	0	0
Total	439	98	22

Source: Adapted from Carbon Catalog (2010).

The total amount and specific mix of offset projects will depend on the price of CO_2 and the relative costs of implementing and managing the individual offset projects. With about 40 percent of the 303 million ha (749 million acres) of U.S. forest land owned by nonindustrial private forest (NIPF) landowners (Smith et al. 2009), assessing the potential of NIPF lands to sell carbon offsets vis-à-vis other options is crucial to developing appropriate carbon offset and forestry policies and programs. Therefore, the objective of this paper is to review the available literature to assess the viability of U.S. forest offset projects relative to alternative carbon offset options to develop a better understanding of how NIPF landowners will respond should Congress pass carbon cap-and-trade legislation. Although a relatively large literature has emerged over the past two decades concerning the costs for a wide variety of carbon offset projects, most of the analyses are sector wide and few, if any, have been directed at the circumstances facing NIPF landowners. In this paper, we review the larger literature on cost of carbon offset projects and make inferences about NIPF lands.

Forestry-Based Carbon Offset Projects

Forests sequester carbon in their biomass through photosynthesis in which CO_2 is absorbed from the atmosphere and stored in roots, stems, leaves, and branches of trees; understory plants; floor litter; and soils. Additional carbon can be sequestered by forests through afforestation, reforestation, or changes in forest management. Afforestation consists of planting trees on previously nonforested lands (e.g., conversion of marginal cropland to trees), whereas reforestation is the replanting of trees on previously forested lands, excluding the planting of trees immediately after timber harvests (USEPA 2009a). Forest management offset projects modify existing forestry practices to enhance carbon storage over time. Examples include lengthening the harvest-regeneration cycle, increasing management intensity, fire control, fertilization, altering stocking densities, choosing alternative tree species, reducing dead biomass removal, reducing harvest intensity, and adopting low-impact logging (Adams et al. 1999, Im et al. 2007, Shaikh et al. 2007, Sohngen and Mendelsohn 2003, Stainback and Alavalapati 2002, USEPA 2009a). Agroforestry[2] systems can also be used to increase carbon sequestration on agricultural lands (Schoeneberger 2005).

Figure 4-1 illustrates Gorte and Ramseur's (2008) estimates from EPA's (2005: C4 p. 21) use of the FASOM-GHG model (see chapter 2) (Adams et al. 2010) of the relative amounts of GHG mitigation from afforestation and improved forest management that may occur at various CO_2e prices. At a CO_2e price of $50 ($15) per metric ton (mmt), more than 800 (100) mmt of CO_2e could be sequestered through afforestation activities, and approximately 380 (210) mmt through improved forest management activities. Table 4-2 shows the amount of land (and percentage of total U.S. agricultural, rangeland, and private forest land) required to sequester 100 to 800 mmt of CO_2e. The values in table 4-2 are based on average rates of sequestration from afforesting nonforest land or through changes in forest management on private forest lands from Lewandrowski et al. (2004) and USEPA (2005) and land use data from the USDA Natural Resources Inventory (USDA NRCS 2003).

The United States currently has a total of 729 million ha of undeveloped lands that could be used for forest based carbon sequestration (149 million ha of crop and pastureland, 164 million ha of rangeland, and 164 million ha of private forestland). Afforesting enough land to sequester 100 to 800 mmt C would require planting trees on 25 to 203 million ha of land, 13 to 103 percent of all cropland or 53 to 430 percent of all rangeland. This would represent an increase of private forest land in

---

[2] Agroforestry is a joint forest production system whereby land, labor, and capital inputs are combined to produce trees and agricultural crops (or livestock) on the same unit of land (Mercer and Pattanayak 2003).

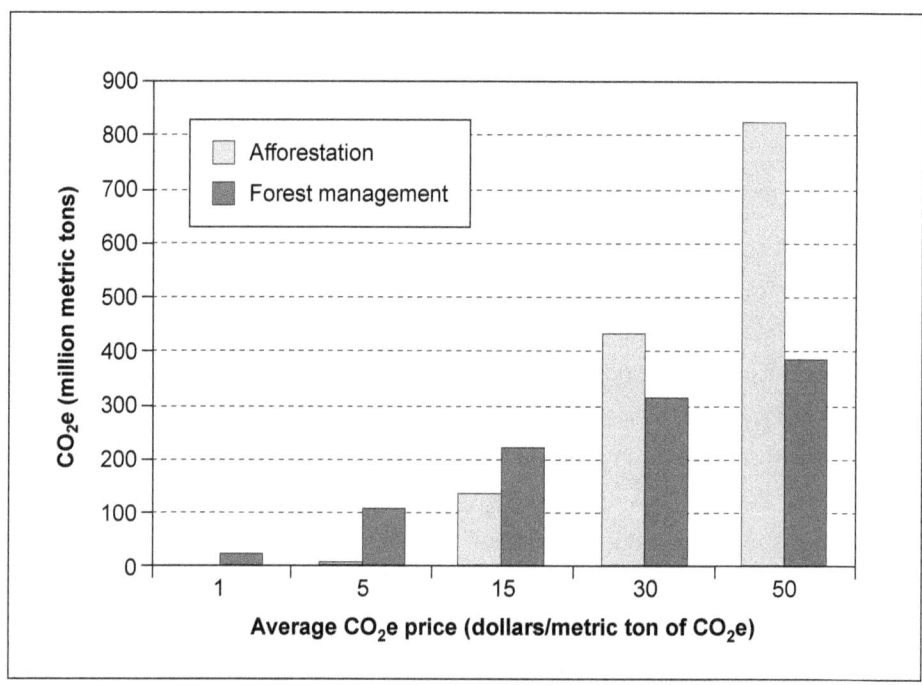

Figure 4-1—Estimated U.S. greenhouse gas mitigation totals for afforestation and forest management: annualized averages, 2010–2110. Source: EPA 2005, Gorte and Ramseur 2008.

Table 4-2—Land required for carbon sequestration by afforestation and management

Carbon offset strategy/land area	Sequestered carbon (million metric tons)		
	100	400	800
Afforestation:			
Land required (million ha)	25.4	101.6	203.2
U.S. agricultural land (percent)	12.9	51.7	103.5
U.S. range land (percent)	53.6	214.5	429.0
U.S. private forest land (percent)	15.5	61.8	123.6
Forest management:			
Land required (million ha)	33.8	135.1	270.1
U.S. private forest land (percent)	20.5	82.2	164.4

Sources: Lewandrowski et al. 2004, USEPA 2005, USDA NRCS 2003.

the United States by 15 to 124 percent, an unprecedented change in land use in the United States for which the economic and ecological repercussions are unknown. The most recent report on tree planting on all lands in the United States showed 1.1 million ha (2.6 million acres) planted in 1997 (Gorte 2009b). Using changes in forest management to sequester 100 to 800 mmt C would require changes in management on 83 to 664 percent of all private forest lands in the United States.

Avoiding emissions by preventing the conversion of forest to nonforest land may also earn offset credits under several of the cap-and-trade proposals. Dedicating lands to continuous forest cover can be established through processes such as long-term conservation easements or transferring forest ownership to the government (CAR 2009). However, the challenge for avoided deforestation projects lies in establishing that forest-land use was threatened by logging or clearing. Compared to reforestation and afforestation, few studies have examined carbon sequestration through avoided deforestation in the United States (Baral and Guha 2004, Langpap and Kim 2010).

Factors Influencing Forestry Offset Projects

A number of factors influence the potential for forest offset projects to reduce GHG emissions including tree species and site characteristics, management practices, longevity of wood products and disposal methods, opportunity costs of land, discount rates, and forest and agricultural prices. These factors result in wide variation in generated GHG mitigation benefits as well as per-unit costs of forest-based GHG reduction.

The ability to sequester carbon in plants and forest soils differs across biomes as shown in figure 4-2. Hardwood (e.g., oak-hickory), softwood (e.g., southern pine, Douglas-fir), and mixed pine-hardwood (e.g., oak-pine) temperate forests are the

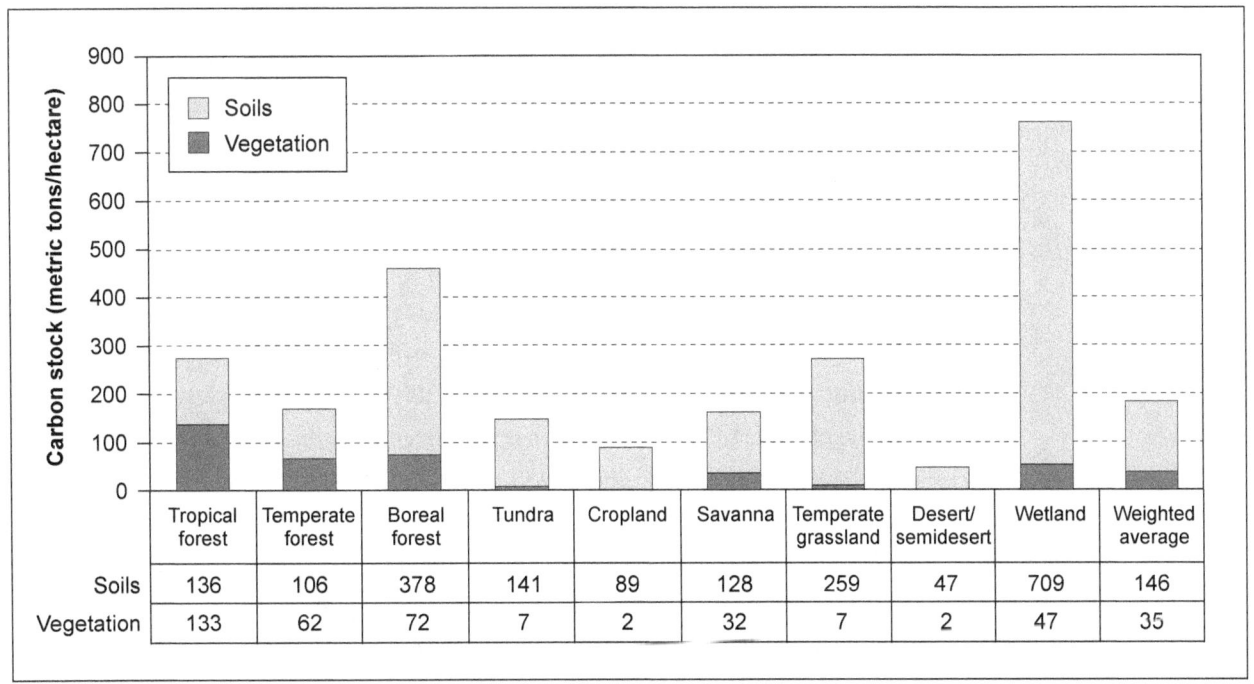

	Tropical forest	Temperate forest	Boreal forest	Tundra	Cropland	Savanna	Temperate grassland	Desert/ semidesert	Wetland	Weighted average
Soils	136	106	378	141	89	128	259	47	709	146
Vegetation	133	62	72	7	2	32	7	2	47	35

Figure 4-2—Variation in average carbon stock in different biomes. Source: IPCC 2005. Adapted from Gorte (2009a).

predominant forest types in the contiguous United States. The carbon sequestered in these types of forests is lower than tropical forests found in the United States (e.g., Hawaii, Puerto Rico, Virgin Islands, Guam, Samoa) and boreal forests in Alaska and Canada. Within a biome, carbon sequestration potential differs across species, climate, soil type, forest age, location, and previous land use. Figure 4-3 illustrates the average distribution of carbon in different components of the forest. Forest soils and vegetation account for most of the carbon accumulated in forests, 59 and 31 percent, respectively.

Rates of carbon sequestration vary widely across the United States, ranging from 2.0 to 10.3 t/ha (0.9 to 4.6 tons/acre) per year (Stavins and Richards 2005). Fast-growing, long-lived tree species would be preferred for carbon sequestration and storage. However, few species have both characteristics. The

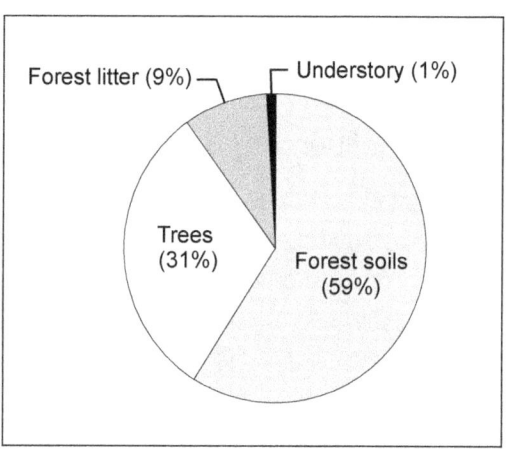

Figure 4-3—Average carbon distribution among forest components. Source: Birdsey 1992.

choice is generally between younger, faster growing trees and older, slower growing stands. Whether slower growing trees can potentially store more carbon over time owing to longer lives depends on a number of factors including geography, site characteristics, management and harvesting regimes, and how the harvested wood is used. For example, Stavins and Richards (2005) showed that carbon sequestration in a southern loblolly pine (*Pinus taeda* L.) plantation peaks at 10.1 t/ha (4.5 tons/ acre) per year and then declines rapidly and becomes insignificant after 70 years. Ponderosa pine (*Pinus ponderosa* C. Lawson) grown in the Mountain States sequesters carbon at a more gradual rate peaking at about 6.7 t/ha (3 tons/acre) per year in year 65 and then declining slowly over the next 100 years. Although ponderosa pine may sequester substantially more carbon than loblolly over the lifetime of the plantation, the carbon uptake occurs much sooner with loblolly. Therefore, species and site choice will depend on the desired timeframe for sequestering carbon.

Within a geographic region, site characteristics play a large role in carbon sequestration potential. Generally, all other things equal, sites with higher site indices will sequester more carbon than those with lower site indices. In addition, less management effort is usually required in higher quality sites resulting in lower marginal costs making the offset projects more lucrative to the landowner. However, species differ in their nutrient needs and growth attributes so that a site that is high quality for one species may not be for other species.

Management Practices

Although most agree that creating new forests through reforestation or afforestation or avoiding the deforestation of existing forests are the most direct and least controversial methods for increasing forest carbon sequestration, many believe that there are relatively few opportunities for large-scale reforestation in the United States (Gorte 2009b, Plantinga et al. 1999, Ray et al. 2009). Because the carbon storage potential of forests depends, in large part, on how the forest is managed, a considerable debate has emerged recently concerning the feasibility of increasing carbon sequestration and storage through changes in forest management.

Forest management practices influence biomass levels, growth patterns of trees, and structure and composition of soils and understory, all of which directly impact carbon sequestration and storage. Actions like thinning (removing some trees to increase growth of desired species) and release (killing competing vegetation chemically or mechanically) may increase carbon sequestration rates by reducing competition for light and nutrients (Gorte 2009a). Newell and Stavins (2000), however, predicted that periodic harvests may increase sequestration costs. Others argued that thinning or release cuts selectively increase total carbon storage in young stands with severe competition, but in other cases, the carbon storage is simply redistributed to fewer large trees (Smith et al. 1997).

The impact of prescribed fire on carbon sequestration remains unclear. In the near term, prescribed fires lead to increased carbon emissions in the atmosphere. Low-intensity prescribed fire usually results in little change in soil carbon, but intense prescribed fire or wildfire can result in significant soil carbon loss (Johnson 1992). Empirical evidence, however, suggests that prescribed burning reduces both the risk of wildfire and the intensity of wildfires, both of which could reduce carbon emissions (Gorte 2009a, Mercer et al. 2007).

It is possible that harvesting timber from mature forests could increase carbon sequestration as young trees grow faster and sequester carbon faster than older trees. However, this depends on how the harvested wood is utilized and the harvesting methods used. Long-lived products like furniture or construction materials can sequester carbon over long periods. However, if the wood is burned as fuel or converted into pulp, paper, or other short-lived products, the carbon will be returned to the atmosphere relatively quickly. Therefore, Perez-Garcia et al. (2005) argued that intensive management, short rotations, and substituting the harvested wood for more energy-intensive building materials may result in a substantial net reduction in atmospheric carbon. Most of the reduction is due to increasing the rate of product substitution rather than by storing more carbon in trees.

If product substitution does not occur, it appears that "passive" forest management may sequester more carbon than intensive forest management because of interactions among storage and uptake rates, basic production ecology, conversion efficiencies from trees to wood products, fuel used for harvesting and management operations and transporting wood to mills and markets, and the dynamics of long-term storage (Perez-Garcia et al. 2005, Ray et al. 2009). Others suggested that harvesting may actually increase the cost of forest-based carbon sequestration (Newell and Stavins 2000, Plantinga and Mauldin 2001, Plantinga et al. 1999). Additional studies are required to determine the conditions under which forest harvesting leads to a net increase or decrease in carbon sequestration.

Product Longevity

The amount of carbon stored in wood products is determined by the type of product and how the products are used. For example, wood used for residential construction can have a usable life of 100 years, whereas paper products generally have a usable life of less than a year (Skog and Nicholson 1998). In addition to the carbon stored (and eventually released) in wood products, a full accounting should include carbon emissions produced during primary and secondary processing, maintenance, repairs, and disposal. Figure 4-4 illustrates Ingerson's (2009) estimate of carbon loss from harvesting wood for a variety of uses. Solid wood products are expected to lose almost 99 percent of the carbon in standing trees within 100 years. However, Ingerson fails to account for the "substitution benefit" of wood compared to petroleum-based products. Although accounting for carbon in products results in a modest increase in carbon storage, including the substitution for more energy-intensive products such as steel or concrete dramatically increases the amount of avoided carbon emissions (Baral and Guha 2004, CORRIM 2009).

Figure 4-5 depicts carbon storage in forests in terms of short- and long-lived products, displacement of fossil fuels, and harvesting and processing emissions for an 80-year rotation. Although the substitution effect is initially about the same as the carbon stored in wood products, over time, the substitution effect dominates. Estimating the total carbon impact of wood products requires data on (1) GHG (carbon) loss occurring at each step in the processing chain from tree harvest to final wood product conversion; (2) carbon emissions resulting from machinery use, transportation, transformation into various products, customer delivery, and disposal into landfills; and (3) carbon savings owing to substitution of wood for

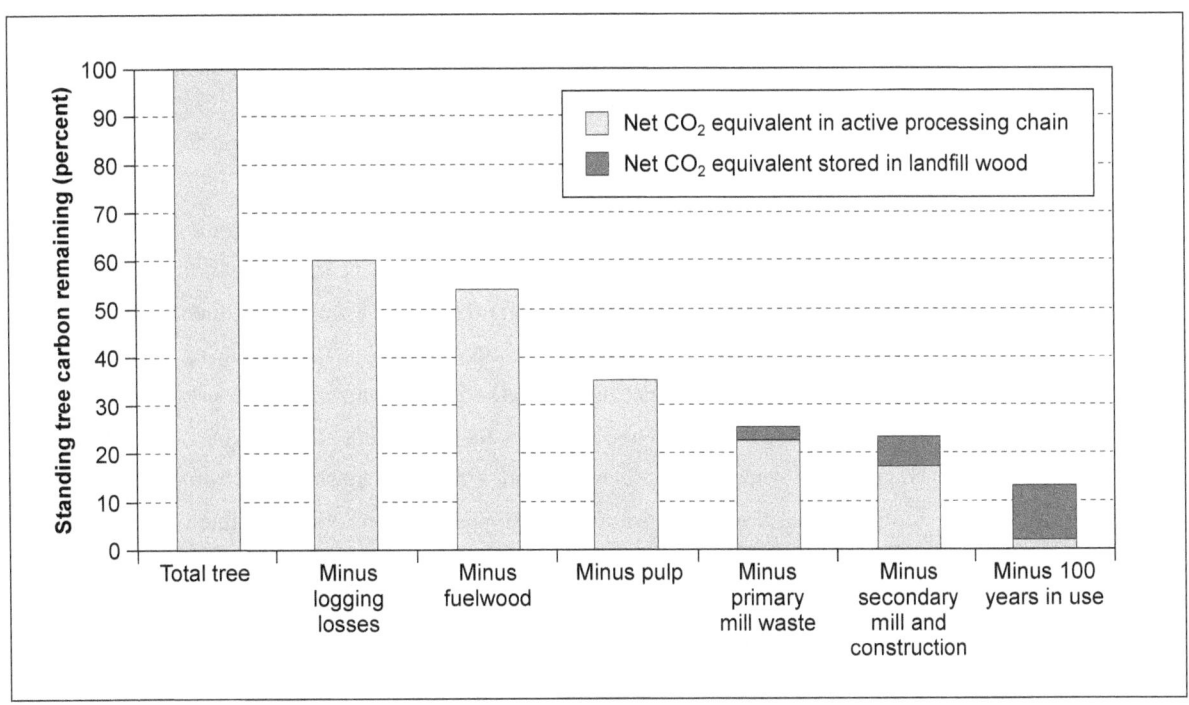

Figure 4-4—Carbon loss from standing trees to products. Source: Ingerson 2009.

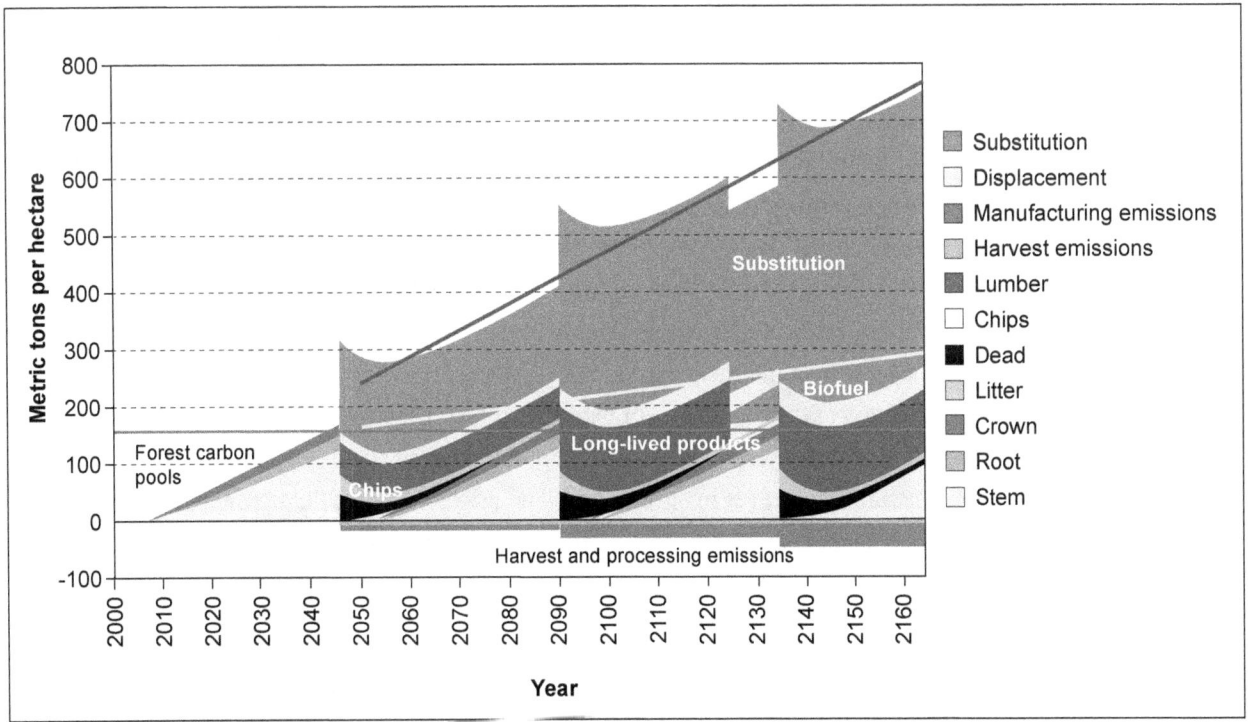

Figure 4-5—Carbon in the forest and product pools, including substitution for concrete construction, for 80-year rotation.
Source: CORRIM 2009.

high-carbon products such as concrete and steel. However, product substitution is not likely to qualify as offsets in cap-and-trade programs in which steel, cement, and petroleum are all regulated sectors.

Opportunity Cost of Land and Discount Rates

The opportunity cost of land, the forgone monetary return from the land in its next best use, is the most important factor in the large variation in the cost of forest offset projects (Stavins and Richards 2005). A number of factors influence the opportunity cost of the land including location, fertility, existing use, climate, market availability, and estimation method. Estimates of the opportunity cost of land have varied from \$0 to \$8,400 per ha (Moulton and Richards 1990, New York State 1991). Some studies factor in anticipated increases in agricultural land prices along with carbon sequestration program expansions and associated reductions in nonforestry land use. The resulting opportunity costs of land range between \$274 to \$5,070 (Richards et al. 1993) and \$116 to \$6,170 per ha (Richards 1997). Typically, carbon sequestration cost estimates in studies that incorporate opportunity costs of land tend to be 2 to 3.5 times higher than others (Manley et al. 2005) and when included, increase total costs by about \$25/t of CO_2 equivalents (CO_2e),[3] on average (van Kooten and Sohngen 2007).

Stavins and Richards (2005) provided a useful summary of the impact of including opportunity costs in three estimation methods: (1) bottom-up engineering cost studies, (2) optimization models that analyze behavioral responses in the forest and agricultural sectors, and (3) econometric analyses of the revealed preferences of landowners for allocating their land to forestry or other alternative land uses. Engineering studies tend to use average of opportunity costs (e.g., average rents for agricultural lands) to estimate the forgone profits from alternative land uses and combine these with tree planting costs to produce a total cost estimate. Sectoral optimization studies use the same basic approach as the engineering studies but also include indirect costs of carbon sequestration programs between different sectors of the economy. For example, if land is converted from agriculture to forest owing to the carbon offset program, the lowered supply of agriculture land may result in increases in agricultural land prices. This would provide incentives for landowners to convert non-carbon-project forest land to agriculture producing a "leakage" of the sequestered carbon.

---

[3] 1 \$US/t C = 3.67 \$US/t CO_2e.

Figure 4-6 illustrates the effect of including the direct costs of land in estimating the impact of prices on forest offset projects. The sector optimization studies (McCarl and Schneider 2001, USEPA 2005) account for both direct costs and the indirect effects on land and other markets, whereas the engineering studies (Moulton and Richards 1990, Richards 1997) only include direct costs. As a result, the engineering studies find significantly greater sequestration at all prices.

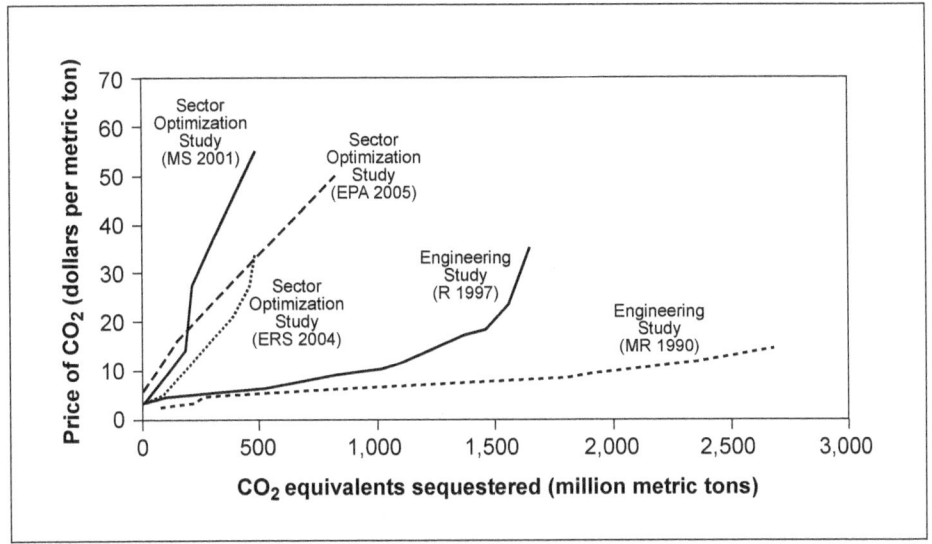

Figure 4-6—Estimates of the annual carbon sequestration through afforestation in the United States at different carbon dioxide (CO_2) prices. EPA 2005 = USEPA 2005; ERS 2004 = Lewandrowski et al. 2004; MS 2001 = McCarl and Schneider 2001; R 1997 = Richards 1997; MR 1990 = Moulton and Richards 1990. Source: CBO 2007.

In contrast, econometric studies are based on actual landowner responses to market conditions. These are also referred to as revealed preference approaches because they use the behavior of landowners to reveal their preferences between different land uses given changes in relative forest and agricultural prices. This approach allows researchers to include a number of additional indirect factors that may influence carbon offset project costs such as uncertainty of the irreversibility of some land use changes, option values, nonmarket benefits (e.g., recreation, aesthetics, biodiversity), liquidity constraints, and other benefits or costs associated with alternative land uses (Stavins and Richards 2005). Typically econometric studies tend to produce higher cost estimates than the engineering or sectoral optimization approaches.

The discount rate also affects the relative cost of offset projects because of the long time horizons for carbon sequestration (Stavins and Richards 2005). The discount rate is simply a particular rate of interest used to bring past or future costs or benefits to the present. Discount rates affect cost estimates by producing changes in (1) the present value of economic variables such as the value of forgone future revenues from harvests, (2) the optimal rotation or harvest age, and (3) the present value of future carbon flows (Stavins and Richards 2005).

One difficulty in comparing forest offset cost estimates is the wide variation in discount rates. For example, a discount rate of 10 percent was used by Moulton and Richards (1990), New York State (1991), and Adams et al. (1993); 7 percent by Creyts et al. (2007); 5 percent by Stavins (1999) and Plantinga et al. (1999); and 4 percent by Parks and Hardie (1995) and Alig et al. (1997). Probably the best approach is to apply a range of discount rates (e.g., 0, 2, 5, and 8 percent) to determine the sensitivity of the results to the discount rate (Richards 1997). For example, Richards et al. (1993) found that increasing the discount rate from 3 to 7 percent doubled the marginal cost of sequestration. However, van Kooten and Sohngen (2007) found that the discount rate had no statistically significant influence on financial costs in their meta-analysis of 68 forest offset cost studies. They attribute this to the fact that most of the financial costs occur in the earliest years of forestry projects, reducing the importance of discounting.

Price Effects of Forest and Agricultural Products

As landowners can switch between agriculture and forest land uses in search of higher profits, changes in prices for agricultural and forest products play an important role in evaluating the potential role and competitiveness of forest offsets. However, the direction of the effect is not always simple or as one might expect. For example, an anticipated increase in future forest product prices compared to agriculture reduces the opportunity cost of forest land and in turn reduces the marginal cost of carbon sequestration and vice versa. Higher forest prices, however, may also produce more frequent harvesting, which may increase sequestration costs (Stavins and Richards 2005). Furthermore, Adams et al. (1993) showed that large-scale conversion of agricultural lands to forestry would produce higher agricultural prices relative to timber prices, which in turn would incentivize landowners to convert their forest land back to agriculture. Newell and Stavins (2000) found that increases in forest land area owing to carbon sequestration subsidies do not necessarily result in continuous increases in carbon storage if the reduced supply of agricultural lands results in higher agricultural prices. They also found that higher agricultural product prices lead to a substantial amount of deforestation, which would at least partially negate the impact of policies subsidizing forest offset projects.

Transaction Costs

Transaction costs of offset projects typically include consulting fees, search and feasibility fees, planning, project documentation, monitoring and verification fees, process determination expenses, insurance charges, and negotiation fees. The transaction flow for a typical forest offset project is illustrated in figure 4-7. Economies of scale hold for transaction costs associated with offset projects because fixed costs (e.g., administrative fees) remain constant, while variable costs decrease with increasing project size (Mooney et al. 2004). As a result, smaller NIPF landowners generally will have to enter into contracts with aggregators to remain competitive with larger projects.

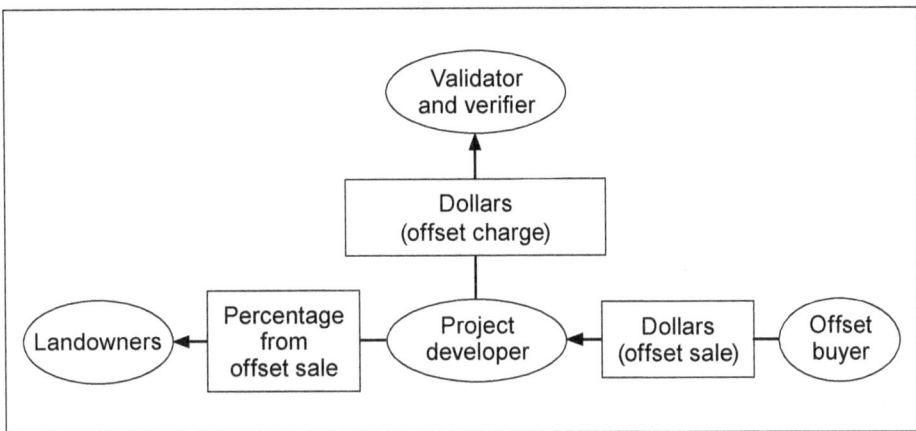

Figure 4-7—Transaction flow for typical forest offset project. Source: Haller and Thoumi 2009.

Galik et al. (2009a) analyzed a hypothetical forest management offset project under multiple accounting methodologies or protocols for a period of 100 years and estimated median transaction costs to be $1.63/t CO_2e sequestered across different regions, forest types, project sizes, and rotation extension lengths. They found that forest offset transaction costs differ by protocol and tend to decrease with project size and length of rotation and that transaction costs can be significant for small forest management offset projects. They conclude that although transaction costs are important, they appear to be less of a factor than the actual accounting scheme (e.g., baselines, leakage, permanence) under which the project is operating. Furthermore, the rules and regulations that would operate under a federal compulsory carbon market regime are still uncertain. Methodology standardization and greater institutional infrastructure will tend to reduce transaction costs and investor uncertainty, whereas more stringent and complex rules and regulations could increase these costs.

Cost Comparison of Offset Projects
Forest Offsets

The many issues, underlying assumptions, and factors discussed above result in a wide variation in the cost estimates of forest offset projects. For example, in a meta-regression analysis of 68 studies, van Kooten and Sohngen (2007) found that total costs ranged from about \$2 to \$280/t CO_2e. European projects were the most expensive, \$48 to \$280/t CO_2e, with costs typically \$212 higher than elsewhere for similar projects.[4] Tropical projects in developing countries, where land and labor costs are low and tree growth rates rapid, are the least expensive (\$2 to \$35/t CO_2e), and U.S. based forest offset projects are competitive ranging from about \$2 to \$77/t CO_2e (van Kooten and Sohngen 2007).

Scientists at Resources for the Future have developed a forest carbon index to analyze the potential of "every piece of land on Earth" to sequester and store carbon in forests (Deveny et al. 2009). The index combines geographic spatial data with forest market data (prices, quantities, revenues), opportunity costs, carbon sequestration potential, investment risks, and cost for government to rank areas and nations according to their likelihood for generating and selling forest carbon credits. First, Deveny et al. (2009) estimated the average cost for forest carbon sequestration for all sources in all countries assuming a market price for carbon less than \$20/t CO_2e. Average forest carbon offset costs range from \$2.11/t CO_2e in the Congo to \$18.95/t CO_2e in Lichtenstein. The United States has the 25th lowest cost (\$7.64 t CO_2e) out of 109 countries studied.

The Forest Carbon Index (FCI) combines each country's carbon offset profit potential (based on average cost and available lands) with risk factors to rank the best places to invest in forest carbon on a 100-point scale. Brazil is the top ranked country (FCI = 100) owing to a combination of high profit potential and low risk. The remaining top 10 countries are Peru (FCI = 95), Bolivia (94), Columbia (93), Indonesia (92), Republic of Congo (92), Russia (91), Central African Republic (91), Democratic Republic of Congo (91), Gabon (91), and Guyana (91). The United States ranks 30th with an FCI of 85. Other temperate zone countries include Canada, ranked 15th with an FCI of 89, and Sweden, Finland, and Australia, all with FCIs of 81 and ranked 40th in the world. Deveny et al. (2009) concluded that 10 countries (Democratic Republic of Congo, Brazil, Angola, Central African Republic, Republic of Congo, Bolivia, Peru, Gabon, Russia, and Cameron) account for 70 percent of the global profit potential for selling carbon credits.

---

[4] Van Kooten and Sohngren (2007) speculated that the lack of competitiveness of forest offsets in Europe may explain why Europe has opposed using biological sinks to offset carbon emissions.

Tables 4-3 and 4-4 present cost estimates from 15 studies of 11 U.S. afforestation offset projects and four forest management offset projects, respectively. Further details on the individual studies can be found in appendix 1.

Costs for afforestation projects ranged from \$1.29 to \$70.92/t CO_2e with an average of \$16.78/t CO_2e and median of \$9.89/t CO_2e. Although the range of cost estimates for forest management projects was smaller (\$12.16 to \$49.35/t CO_2e), the average and median costs were higher at \$27.65/t CO_2e and \$24.54/t CO_2e, respectively. This suggests that, in general, afforestation projects will be more cost competitive than forest management projects. Likewise, Sohngen and Mendelsohn (2003) predicted that sequestration costs would be higher for forest management projects compared to afforestation projects.

Table 4-3—Costs (in 2005 dollars) of removing atmospheric carbon dioxide (CO_2) through afforestation offset projects

Forest carbon sink studies	Total carbon	Total area	Cost Per hectare	Cost Per t carbon	Cost Per t CO_2	Estimation method	Discount rate	Study location
	Million metric tons	*Million hectares*	- - - - - - - *Dollars* - - - - - - -				*Percent*	
Adams et al. 1993	140	59	442.28	73.2	19.94	Sector optimization model	10	U.S.
Adams et al. 1999	2023.08	145.6	401.52	29.16	7.95	Sector optimization model	4	U.S.
Baral and Guha 2004	316.75	1	18 602.34	63.3	17.25	Bottom-up approach	N/A	U.S. South
Callaway and McCarl 1996	119.32	29.62	143.39	34.09	9.29	A modified version of the Agricultural Sector Model (ASM)	10	U.S.
Dixon et al. 1994	5.98	0.03	180.72	4.73	1.29	Bottom-up calculations	4	Oregon, Utah
McCarl and Callaway 1995	243.88	47.39	383.74	72.36	19.72	Sector optimization through modified ASM	10	U.S.
New York State 1991	0.5	0.8	17.33	29.51	8.04	Bottom-up calculations	10	New York
Parks and Hardie 1995	29.96	6.58	967.26	260.29	70.92	Engineering cost curve approach	4	U.S.
Plantinga and Mauldin 2001	41.55	0.28	5457.4	36.28	9.89	Econometric estimation method	5	Maine, South Carolina, Wisconsin
Plantinga et al. 1999	12.8	0.19	4596.33	67.61	18.42	Econometric estimation method	5	Maine, South Carolina, Wisconsin
Richards et al. 1993	42 903	86.4	3446.72	6.94	1.89	Engineering cost curve approach	5	U.S.
Average	4167	34.26	3149	61.59	16.78			
Median	119.32	6.58	442.28	36.28	9.89			
Minimum	0.50	0.03	17.33	4.73	1.29			
Maximum	42 903	145.60	18 602	260.29	70.92			
Standard deviation	12 262	45.13	5233	67.06	18.27			

Source: Adapted from van Kooten and Sohngen (2007).

Table 4-4—Costs (in 2005 dollars) of removing atmospheric carbon dioxide (CO_2) through forest management offset projects

Forest carbon sink studies	Total carbon	Total area	Cost			Estimation method	Discount rate	Study location
			Per hectare	Per t carbon	Per t CO_2			
	Million metric tons	Million hectares	- - - - - - - Dollars - - - - - - -				Percent	
Huang and Kronrad 2001	0.06	0	838.78	44.63	12.16	Average cost of carbon storage calculated through optimal rotation age, different interest rate, and site index	2.5–15	Texas
Sohngen and Brown 2006	2.28	0.22	1921.95	130	35.42	Engineering cost curve approach	6	U.S. South and West
Sohngen and Haynes 1997	29	198	7.34	50.1	13.65	Linking forest fire mortality model to forest inventory model	5	Contiguous U.S.
Newell and Stavins 2000	7.66	2.07	699.79	181.13	49.35	Econometric estimation approach	5	Arkansas, Louisiana, and Mississippi
Average	9.75	50.07	866.97	101.47	27.65			
Median	7.66	2.07	838.78	101.47	24.54			
Minimum	0.06	0.00	7.34	44.63	12.16			
Maximum	29.00	198.00	1921.95	181.13	49.35			
Standard deviation	11.45	85.41	685.69	57.07	15.55			

Source: Adapted from van Kooten and Sohngen (2007). The cost estimates in the table are drawn from meta regression analysis. As studies in their sample provided multiple estimates of one or more projects or regions, the averaged values across a study were calculated by van Kooten and Sohngen (2007).

Although most studies calculate average costs (total project cost divided by the number of units of CO_2e sequestered), a more useful measure is the marginal cost (the cost of reducing one additional unit of CO_2e). In contrast to average costs, marginal cost analysis allows the cost to change with market situations. For example, Stavins and Richards (2005) used a normalized set of forest studies and a 5-percent discount rate to find that 300 million metric tons of carbon can be sequestered at a carbon price of $7.50 to $22.50 per metric ton of CO_2e but that as the amount of sequestration increases, e.g., to 600 million metric tons per year, the marginal cost increases significantly.

Agriculture Offsets

Agricultural projects can produce carbon credits in two ways, increasing the amount of carbon sequestered in the soil or implementing management practices that reduce the amount of agricultural emissions. Activities that increase carbon in the soil include altered tillage practices (e.g., low tillage or no tillage), planting winter cover crops and other practices to increase biomass returned to the soil, rotating crops, converting monocultures to diverse systems, reducing carbon emissions from organic soils, and establishing perennial vegetation on steep slopes. Projects aimed at reducing emissions include reducing nitrogen fertilizer use or increasing nitrogen use efficiency, changing frequency and duration of flooding of rice paddies, and, reducing GHG emissions from manure and effluents owing to changes in animal management practices, including dietary modifications.

Figure 4-8 provides the Congressional Budget Office (CBO 2007) comparison of estimates of annual per-unit sequestration costs for agricultural sector offset projects by McCarl (2007), USEPA (2005), McCarl and Schneider (2001), and Lewandrowski et al. (2004). All the studies in figure 4-8 used sector optimization techniques and include both direct and indirect (opportunity) costs in their analyses.

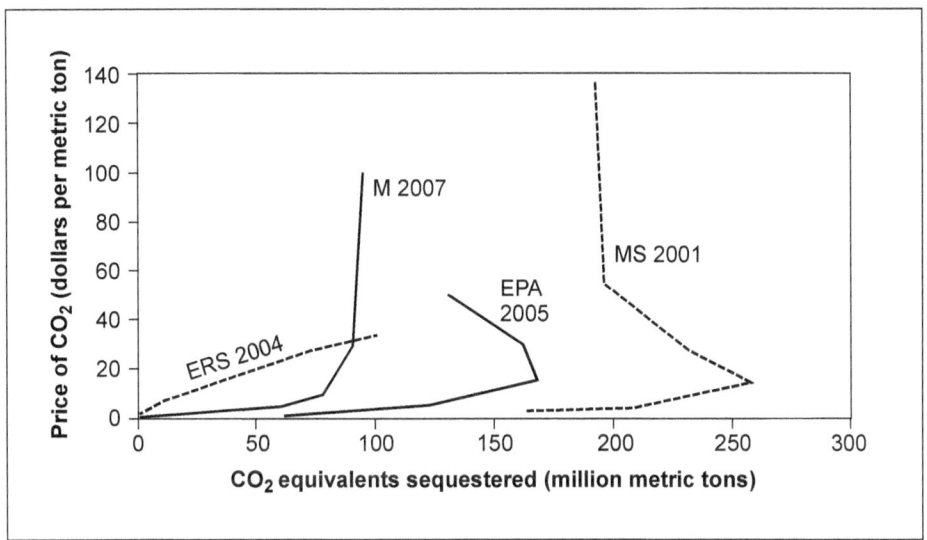

Figure 4-8—Estimates of the annual carbon sequestration from U.S. croplands at different carbon dioxide (CO_2) prices. M 2007 = McCarl 2007; EPA 2005 = USEPA 2005; ERS 2004 = Lewand-rowski et al. 2004; MS 2001 = McCarl and Schneider 2001. Source: CBO 2007.

Initially when the price of CO_2 is below $15, sequestration and associated costs are quite low. However, costs rise rapidly as more agricultural sequestration takes place and more valuable and productive land would be required to increase seques-tration, resulting in a backward bending supply curve. The upper bound of carbon sequestration potential from agriculture is reached at a price between $15 and $20/t CO_2. However, the projected amount of sequestration at different prices differs widely among the studies in figure 4-8. For example, at a price of $10/t CO_2, the projected amount of carbon sequestered ranges from about 10 to 260 million metric tons, and at a price of about $40, estimated carbon sequestered ranges from 80 to 190 million metric tons.

Table 4-5 provides a summary of studies providing estimates of sequestration costs for a variety of agricultural offset projects including no tillage, conservation tillage, shifting from crops to grass, and shifting to continuous cropping. Costs range from about \$2 to \$215/t CO_2e with an average of \$64.93 and median of \$55.90/t CO_2e.

Table 4-5—Costs (in 2005 dollars) of removing atmospheric carbon dioxide CO_2 through agricultural offset projects

| Type of offset project | Cost | | Estimation method | Discount rate | Study location | Citations |
	Range	Average				
	- - - - - $/t CO_2e$ - - - - -			*Percent*		
No tillage agriculture	68.62 to 360.98	214.80	Econometric model linked with a biophysical model to get changes in soil carbon	N/A	Iowa	Pautsch et al. 2001
Reduced tillage	3.08 to 154.46	78.77	The ASMGHG model—a market and spatial equilibrium sector optimization model	4	U.S.	McCarl and Schneider 2001
No till versus conventional tillage	1.68 to 153.93	77.81	Experimental yields, input rates, field operations, and prices simulate a distribution of net returns for production systems	N/A	Kansas	Pendell et al. 2006
Conservation tillage	3.31 to 41.43	22.37	The U.S. Agricultural Sector Model (USMP), a spatial and market equilibrium model	5	U.S.	Lewandrowksi et al. 2004
No tillage	0.53 to 2.74	1.64	Meta-regression analyses of 52 studies of net returns and 51 studies of carbon soil uptake	N/A	South	Manley et al. 2005
No tillage	44.31 to 113.00	78.66	Meta-regression analyses of 52 studies of net returns and 51 studies of carbon soil uptake	N/A	Prairies	Manley et al. 2005
No tillage	25.27 to 42.7	33.99	Meta-regression analyses of 52 studies of net returns and 51 studies of carbon soil uptake	N/A	Corn Belt	Manley et al. 2005
Change from crops to permanent grass	16.95 to 169.39	93.17	Field-level econometric production models combined with crop ecosystem model	N/A	Montana	Antle et al. 2001
Change from crops to permanent grass	3.31 to 41.43	22.37	The U.S. Agricultural Sector Model (USMP), a spatial and market equilibrium model	5	U.S.	Lewandrowksi et al. 2004
Shift to continuous cropping	4.07 to 47.47	25.77	Field-level econometric production models combined with crop ecosystem model	N/A	Montana	Antle et al. 2001
Average		64.93				
Median		55.90				
Minimum		1.64				
Maximum		214.80				
Standard deviation		58.24				

Capture and Storage/Destruction of GHG Emissions

Carbon dioxide capture and storage (CCS) projects capture CO_2 from electric powerplants or industrial sources and place it in long-term storage, e.g., by injecting it underground or storing it in oceans. Another option is to destroy non-CO_2 GHGs through activities like methane management in coal mining and natural gas and petroleum systems, and controlling nitrous oxides and fluorocarbons in production processes. These CCS projects are predicted to potentially reduce 15 to 55 percent of total carbon emissions by the end of the century (Metz et al. 2005) at a cost ranging between \$15 and \$90/t CO_2e (CBO 2007). The variation in the cost of CCS is a result of variation in the types and sizes of proposed CCS projects. For example, Herzog et al. (2003) estimated the cost of coal-based CCS to be \$50 to \$65/t CO_2e compared to gas-based CCS at \$85/t CO_2e.

The CCS technology is still at a very early stage of deployment, with few plants operational worldwide. A major impediment is the energy loss as CCS technology is expected to use somewhere between 10 and 40 percent of the energy produced by the plant (Rochon 2008). As compared to CCS projects, projects that destroy non-CO_2 GHGs tend to have lower costs per unit CO_2e reduction. Creyts et al. (2007) estimated per-unit net cost for these projects to be \$3/t CO_2e compared to \$49 for CCS projects.

Offset Cost Competitiveness

Table 4-6 compares the gross and net costs of GHG mitigation for different offset options. The gross costs are derived from tables 4-3, 4-4, and 4-5. Creyts et al.'s (2007) estimates of net costs in table 4-6 are a weighted average of the opportunities within each sector and reflect the capital, operating, and maintenance costs minus any energy savings associated with abating 1 metric ton of CO_2e per year over a 25-year forecast period using a 7-percent discount rate. Net cost values in parentheses represent options that provide net benefits; i.e., the energy benefits are greater than the costs of the options. All estimates (gross and net) do not include transaction costs, taxes, tariffs, subsidies, or incentives.

Conservation tillage is expected to produce net benefits of \$7/t CO_2e owing to energy savings from reduced use of machinery and fertilizers with reduced- or no-tillage systems. All other biological sequestration sources are expected to produce positive gross and net costs and are expected to be competitive with all other sources except recovery and destruction of non-CO_2 GHGs.

Of the forest offset options, afforesting pasturelands appears to be the most cost-effective option. Afforesting croplands is more costly because of the higher opportunity cost of cropland compared to pastureland. Creyts et al. (2007)

Table 4-6—Cost competitiveness of U.S. offset projects

Type of greenhouse gas offset option	Average gross cost	Average net cost	Potential
			Million metric tons CO_2e
	2005 dollars/ton CO_2e		
Forest offsets:			
Afforestation of cropland	17	39	80
Afforestation of pastureland	17	18	130
Improved forest management	28	23	110
Agricultural offsets:			
Conservation tillage	73	7	80
Winter cover crops	47	27	40
Reducing already-emitted greenhouse gases:			
Industrial carbon capture and storage	38	49	95
Carbon capture and storage in powerplants	38	44	290
Recovery and destruction of non-CO_2 greenhouse gases	n/a	3	255

Source: Creyts et al. (2007) and author compilation.

estimated that 6.9 million ha of pasture land (4.2 percent of all rangeland in the United States) and 5.3 million ha of agricultural lands (2.7 percent of all agricultural land) could be reforested. Reforesting pastureland is expected to sequester 130 million metric tons of CO_2e annually with 50 percent of the carbon reduction occurring in the South owing to the region's rapid carbon uptake, low conversion costs, and relatively low opportunity costs. Afforesting croplands is expected to sequester about 80 million metric tons of CO_2e with the most cost-effective opportunities in the South. The CBO (2007) concluded that a price of $5/t CO_2e would prompt enough tree planting to sequester between 2 and 50 mmt annually and that a price of $50/t CO_2e would increase the amount sequestered to 500 to 800 mmt annually. However, others suggest that the lack of availability of lands for afforestation may limit the usefulness of this option in the United States (Gorte 2009b, Plantinga et al. 1999, Ray et al. 2009).

Along with the per-unit cost estimation, scalability of different types of offset projects also needs to be factored in. Some options like CCS are in the early stages of development and require large upfront costs, making them quite limited in scope for now. In contrast, soil carbon sequestration through agriculture offsets might make their most substantial contribution at low CO_2 prices; however, the total amount of soil carbon that can be sequestered is limited owing to requirements for soil saturation potentials ranging from 25 to 260 mmt (Gorte 2009a).

The Southern United States has many opportunities for additional afforestation as part of climate change mitigation.

Policy Environment

A number of current and proposed policies and programs may influence the competitiveness of forest offset projects. Some of these policies are directed specifically at GHG reduction, whereas others influence offset projects indirectly. Regulatory mechanisms, incentive-based policies, and government support programs may influence the competitiveness of forest carbon offset projects. Regulatory mechanisms include polices that set goals/targets/limits, and compel certain types of behavior. Incentive-based policies provide financial incentives such as cost-shares, tax reductions, subsidies, or grants, and low- or no-interest loans for project financing. Support programs create supportive infrastructure, provide research and development support, and facilitate public educational outreach.

Cap-and-Trade Initiatives

In the absence of federal leadership on climate change, a number of states have taken the initiative to establish regional cap-and-trade regulations. The Regional Greenhouse Gas Initiative (RGGI) is an agreement between 10 Northeastern and Mid-Atlantic States to reduce CO_2 emissions by 10 percent by 2018. The RGGI allows regulated emitters to use offsets to satisfy 3.3 percent of their compliance obligations initially with provisions to allow offset ceilings to rise to 5 percent and 10 percent of total emissions if the 12-month rolling average allowance reaches

$7 and $10 per ton of CO_2 (in 2005 dollars), respectively. Eligible offset projects include landfill methane capture and destruction, afforestation, sulphur fluoride reduction in the electricity sector, avoided agricultural methane emissions, and energy-efficient building projects. The Western Climate Initiative (WCI) is a comprehensive regional effort by the governors of seven U.S. states and premiers of four Canadian provinces to reduce GHG emissions to 15 percent below 2005 levels by 2020 (WCI 2007). Offset projects will be allowed, but the program is still in the planning stages. Nine states and two Canadian provinces are also developing the Midwestern Regional Greenhouse Gas Reduction Accord (MGGRA). The MGGRA is scheduled to start in 2012 and will incorporate a regional cap-and-trade system covering most sectors of the economy with an emissions target of 16 percent below 2005 levels. Offset criteria, eligibility, and implementation remain in the planning stage.

In addition, a number of states are compiling GHG inventories and designing and implementing emission reduction programs. Over 45 states have completed GHG inventories, and more than 38 have developed (or are in the process of developing) state action plans to reduce emissions (USEPA 2009b). Various states have also promulgated different options to deal with GHG emissions. As many as 35 states have some sort of energy portfolio standards, 13 have appliance/equipment standards for energy efficiency, 42 have energy standards for public buildings, and all the states have energy codes for buildings (NCSC et al. 2010). Many states have also formulated alternative fuels vehicle acquisition regulations, whereby states set priorities for purchasing alternative-fuel vehicles for state or local agencies.

Incentive-Based Policies

A number of programs and policies provide financial incentives to encourage the adoption of a wide array of agricultural, forest, and other land uses to enhance the production of specific commodities; encourage the production of ecosystem services; and to promote sustainable production systems. These incentives complicate the analysis of the competitiveness of alternative carbon offset projects as they influence the opportunity costs of alternative land uses and production processes. It is also uncertain how receiving incentive payments from these various programs will affect eligibility to sell carbon offset credits, i.e., whether stacking of payments will be allowed and how they will affect additionality.

Motivated by fears of timber scarcity, state and federal governments first introduced programs providing financial incentives to NIPFs in the 1940s with the goal of increasing timber production and supply for the postwar construction boom. A variety of approaches were tried including cost sharing; technical assistance;

property, estate, and income tax reductions; and preferential capital gains treatment for timber production (Kilgore et al. 2007). Beginning with the 1990 Farm Bill, incentive programs for forest landowners began to shift focus from timber to forest stewardship, conservation, and the production of ecosystem services. Although the objectives have changed over the past 60 years, the policy levers have remained fairly constant emphasizing cost sharing, technical assistance, and tax incentives. Currently there are at least 14 federal programs that encourage private landowners to adopt stewardship practices to enhance ecosystem services through improved forest management, retention of lands in forest or undeveloped uses, protection of soil and water quality, enhancement and preservation of forested wetlands, and wildlife habitat improvement.

Several federal forest and agriculture programs also provide incentives for mitigating GHGs. For example, the Environmental Quality Incentives Program provides cost-share assistance to install GHG mitigating technologies and the Conservation Reserve Program, Wildlife Habitat Incentives Program, and Landowners Incentive Program provide financial assistance to landowners for a variety of conservation goals including carbon sequestration. The Forest Land Enhancement Program promotes additional carbon sequestration as well as other ecosystem benefits through cost-share partnerships with landowners, and the Biomass Crop Assistance Program provides financial assistance to producers that deliver eligible biomass material to designated biomass conversion facilities.

State governments began initiating cost-share programs in the 1970s and 1980s to supplement federal funding. Similar to federal programs, the availability of state management and cost-share assistance funding has fluctuated over the years. The largest programs in terms of payments and area treated have historically been in the South. Although state programs initially focused on timber productivity, over the past 25 years, the focus has expanded beyond timber to promote the retention of agricultural and forestry land uses, protection of riparian areas and wetlands, enhancement of wildlife habitats, and water quality and soil conservation. State programs primarily assist with development of management plans and cost-share assistance to implement stewardship practices. At least 27 states have adopted cost-share assistance programs; these include eight in the South, nine in the Midwest, five in the West, and five in the East. Most state programs prohibit payments from both federal and state sources for the same practice, but a few do allow both sources of funding up to 100 percent of the cost of the project (Greene et al. 2005).

Property taxes have the greatest potential of any state tax to influence land use decisions. All states in the United States assess or tax forest land at preferential rates, either as timberland or as agricultural or unproductive land. The states differ substantially in the approaches they use and the methods by which they apply them. Many states—particularly in the North—use a yield tax approach, in which the forest is divided into land and timber components. The land is taxed annually, but the tax on the timber is deferred until it is harvested. Other states—particularly in the South—use a modified assessment approach, in which rural land is assessed differently from other property. The assessed value may be fixed, calculated using a reduced assessment rate, or based on the land's actual use instead of its "highest and best" use. Still other states use an exemption approach, which removes forest land, timber, or both from the property tax rolls, either permanently or for a set number of years. A number of states use two or more of these approaches (Siegel and Hickman 1989).

Landowners have retained subsidized afforested stands at high rates and well beyond government program life.

Support Programs

A robust system of agricultural and forestry extension support programs at federal and state agencies and universities facilitate knowledge transfers, technology demonstrations, and information sharing to landowners and producers. Extension agents and specialists at land grant universities and government institutes transfer natural resource management (including GHG mitigation information) knowledge to farmers, forest owners, foresters, and other natural resource managers. These programs are also being used for extending knowledge regarding GHG mitigation and offset benefits. Examples include GHG technology demonstration initiatives such as anaerobic digesters, geothermal, and wind power to farmers, local government, communities, industries, landowners, and consumers. The USDA collaborates with private partners to develop pilot projects for testing forest and agriculture GHG sequestration and mitigation technologies and practices.

Effectiveness of Forest Incentives

Although several studies have demonstrated that cost-share payments have increased reforestation and tree planting behavior (Brooks 1985, de Steiguer 1984, Hyberg and Holthausen 1989, Lee et al. 1992), others have concluded that many landowners received cost-share payments for activities they would have done anyway and that the cost-share payments effectively substitute public for private investment (Baughman 2002, Boyd 1984, Brockett and Gerhard 1999, James et al. 1951, Kluender et al. 1999, Zhang and Flick 2001). For example, Cohen (1983) concluded that 30 to 50 percent of all land reforested under cost-sharing agreements would have been planted without the incentives. Others, however, found that the cost-share payments allowed landowners to expand the number of acres they were able to reforest (Bliss and Martin 1990).

Following a thorough review of more than 50 years of literature and conducting focus groups of landowners across the United States, Greene et al. (2005) concluded that most financial incentives actually have little effect on forest owner behavior. In their meta-analysis of 41 econometric studies, Beach et al. (2005) examined the impacts of four categories of factors that influence forest management decisions: market drivers (e.g., prices, costs, and returns to alternative investments), policy variables (tax incentives, cost-sharing, and technical assistance), owner characteristics, and site conditions. Policy variables were the most likely to be significant when included (87 percent), followed by plot/resource conditions (79 percent), owner characteristics (77 percent), and market drivers (73 percent). Both Beach et al. (2005) and Greene et al. (2005) concluded that NIPF owners more often respond to targeted government programs than to market prices or other financial incentives and that three approaches have consistently succeeded in changing forest management decisions by private landowners: technical assistance, cost-share payments, and direct contact with professional foresters or natural resource specialists. For example, as early as 1951, forest landowners were shown to prefer technical assistance over financial or tax incentives (James et al. 1951). More recently, Greene and Blatner (1986), Baughman (2002), and Kilgore and Blinn (2004) found technical assistance to be the most effective way to change forest landowner behavior in both the United States and Canada.

Impacts of Potential Offset Rules and Criteria

The specific rules and regulations for implementing and receiving credit for offset projects will have a major impact on the competitiveness of forest offset projects. Offset criteria like additionality, baselines, permanence, leakage, and stackability, may influence the competitiveness of forest offset projects.

Additionality and Baseline

Eligible offset projects in all existing and proposed cap-and-trade programs are required to sequester carbon that is in addition to what would have been sequestered in the absence of the project. To prove additionality, most cap-and-trade proposals require that projects be able to show that they have sequestered carbon above and beyond an established preproject baseline. Establishing baselines for afforestation projects should not be too difficult. However, establishing baselines for sequestering carbon through changes in forest management practices (e.g., extending rotation ages) is difficult, controversial, and potentially costly. In addition, once the baseline rotation age is established, periodic monitoring and verification will likely be required. This will entail reviews of records, site visits, and independent measurements of the carbon stocks, all of which will incur additional cost (Sohngen and Brown 2006).

The two general approaches for establishing carbon baselines are **business as usual** (BAU) and **base year**. In the BAU approach, a reference case of projected future carbon stocks in the absence of the project is established as the minimum performance standard for selling offset credits. Following project implementation, the actual increases in forest carbon stocks are compared to the reference case. The BAU can be applied to specific projects (**project-specific performance standards**) or to broad project types or economic sectors (**standardized or group performance standards**). The base year approach determines the difference in the carbon stocks from one time (the base year) to another. Any additions to the carbon stock during the period would be eligible. Within these two general approaches, a number of accounting schemes or protocols have recently been suggested to track the mitigation achieved by individual forest management projects.

Galik et al. (2009b) examined the impacts of seven proposed accounting systems on simulated 100-year forest management projects extending the rotation age in loblolly pine stands in South Carolina. One hundred years post project implementation, net sequestration among the seven protocols differed by almost an order of magnitude. The wide variation is due to differences in how protocols address individual carbon pools, baseline, leakage, certainty, and buffers.

The variation in net sequestration has a significant impact on the break-even carbon price, potentially leading to higher project costs than estimated in previous aggregate national analyses (Galik et al. 2009b). Break-even carbon prices range from $10 to almost $200/t CO_2e depending on the accounting method and the range of values for baseline, reversal, leakage, and uncertainty used in the analysis. Sohngen and Brown (2006) also found higher marginal costs and lower potential for carbon offset projects based on extending rotation ages than previous studies.

Permanence of Offsets

The likelihood that sequestered carbon will be permanently removed from the atmosphere is a major issue associated with forest carbon offset projects owing to uncertainties over the possibility of reversal associated with future timber harvests, land use change, and natural disturbances such as wildfire, hurricanes, and pest and disease outbreaks. Permanence also is a major issue for agricultural offsets. For example, Lewandrowski et al. (2004) found that all the carbon stored from 20 years of conservation tillage could be released into the atmosphere within 1 year after returning to conventional tillage practices. Permanence uncertainty puts forest and agricultural offset projects at a significant disadvantage relative to fugitive emissions[5] reduction offsets such as methane digesters.

A variety of approaches to the permanence problem have been suggested. These include developing some sort of insurance or risk pooling mechanism; requiring a version of a conservation easement or deed restriction (Climate Action Reserve, Georgia Forestry Commission, RGGI); developing buffer pools in which a percentage of issued credits are placed in a "savings account" (Chicago Climate Exchange, RGGI, Voluntary Carbon Standard); requiring carbon banks of forest lands managed for carbon sequestration as replacement reserves should reversals occur; and requiring projects to engage in management activities that reduce the risk of wildfire, pests, and diseases.

Other approaches include developing a conversion factor to convert temporary carbon storage into a permanent equivalent exchange rate in term of ton-years (Dutschke 2002, Herzog et al. 2003, Watson et al. 2000). Estimates of exchange rates range from 42 to 150 ton-years of temporary storage to cover 1 permanent ton (van Kooten and Sohngen 2007). Another alternative would be to issue temporary carbon emission reduction credits (TCERs) that would expire after a set period (e.g., 5 to 15 years). Upon expiration of the stipulated period, TCERs would be reissued or substituted for other credits. Others (Bigsby 2009, Chomitz and Lecocq 2004, Marland et al. 2001) have proposed a rental system in which carbon emitters would rent carbon credits from landowners for a set period. The annual rental rate would be the market-determined price of a permanent emission credit multiplied by the discount rate. Bigsby (2009) takes this rental market a step forward by proposing a carbon banking scheme like capital markets, where the bank creates a "carbon pool" from carbon credit deposits by carbon owners who receive annual payments. The banks could then loan the carbon credits to borrowers who pay an annual rental.

---

[5] Fugitive emissions are pollutants released into air from leaks in equipment, pipelines, seals, valves, etc., and not from the usual sources such as chimneys, stacks, and vents.

Leakage

Leakage refers to changes in emissions or sequestration influenced by a project, policy, market, or other entity that occur outside the boundaries of the particular project, policy, market, or other entity. Several forms of leakage are possible. **Internal leakage** refers to changes in nonproject areas (under the same ownership) to make up for changes made in the project area. For example, shortening the timber rotation in one area in response to extending rotation ages in others would be a form of internal leakage. When a project activity produces changes by other owners or entities, **external leakage** has occurred. **Market leakage** occurs when the project changes the availability of market goods that results in changes in emissions. For example, reducing harvest in the United States could result in increasing harvests in other countries to make up for the reduced supply of wood products. Jenkins et al. (2009) described six potential policy approaches for dealing with leakage:

- Improve project monitoring and design.
- Discount offset credits to account for leakage.
- Develop systemwide (e.g., sectoral, regional, national) accounting protocols to track and reconcile leakage.
- Reduce carbon cap to account for leakage.
- Expand scope of eligible activities so that fewer can leak.
- Ignore leakage but acknowledge the potential for error in the system.

Methods for dealing with leakage vary widely between the carbon accounting protocols. Of the protocols analyzed by Galik et al. (2009b), only the Voluntary Carbon Standard (VCS) and Harnessing Farms and Forests (HFF) require quantification of leakage and provide guidelines. Assessing and quantifying leakage were optional under the California Climate Action Registry (CCAR), but the new forestry protocol for the California Action Reserve requires an estimate of leakage (which they sometimes call "secondary effects") for all forestry projects. The state of Maine proposed that RGGI require project lands to be certified against leakage or that harvests meet or exceed average removal rates for the area. The estimated impact of these requirements are deductions from creditable carbon ranging from 0 percent (CCAR, RGGI) to 10 percent (VCS) to 43 percent (HHF) of the annual creditable carbon (Galik et al. 2009b).

Stackability

The issue of stackability involves offset projects that in addition to sequestering carbon produce other ecosystem services for which additional payments are available (e.g., from government incentive programs such as the Conservation Reserve Program [CRP]. For example a reforestation project could produce habitat for wildlife species, enhance biodiversity, reduce soil erosion, and improve water quality while at the same time sequestering carbon. One might imagine that a landowner for such a project would be eligible to receive cost-share payments from government programs, tax credits or incentives, water quality trading credits, or hunting lease payments in addition to selling the carbon offset credits. This would allow landowners to manage an entire portfolio of payments for ecosystem services. Another example would be combining a carbon offset project with forested riparian buffers that generate water quality market credits and understory thinnings to generate endangered species habitat credits (RFF 2009).

The additional revenues associated with stacked payments for multiple ecosystem services could be crucial for enhancing the competitiveness of NIPF offset projects. Based on the available data, Mercer et al. (in press) found that between 2005 and 2007, landowners received annual payments for ecosystem services (PES) of at least $1.9 billion per year for forest-based ecosystem services. In 2007, private forest landowners in the United States received $727 million for wetland mitigation bank credits, $34 million for conservation bank credit, $1.7 million for sales of carbon offsets, $315 million for conservation easements, and $410 million for hunting leases.

Nevertheless, few landowners participate in these programs. For example, wetland mitigation accounted for the largest percentage of forest-based ecosystem service payments, with 38 percent of all payments in 2007. However, these payments were received by only about 173 mitigation banks, about 0.00002 percent of all private forest landowners. In addition, only 5.5 percent of family forest landowners ever received cost-share payments and only 1.8 percent have conservation easements (Butler 2008). As a result, private forest landowners consistently report that forest-based PES have had little impact on changing their behavior (Greene et al. 2005). A large literature has questioned the additionality of forest-based PES in the United States, suggesting that a large portion of government incentive payments paid landowners for what they would have done without the financial incentives (Baughman 2002, Boyd 1984, Brockett and Gerhard 1999, Cohen 1983, James et al. 1951, Kluender et al. 1999, Zhang and Flick 2001).

Existing programs and pending legislation do not fully address stacking of carbon (GHG) payments with other federal or state payments or programs (Olander et al. 2010). Two major issues that would need to be addressed are additionality and double-counting (RFF 2009). The additionality issue becomes more complex with stacked payments, as the landowner would need to insure that all payments (not just the carbon payments) result in additional provision of the targeted ecosystem service. For example, paying for carbon offsets under a cap-and-trade regime

Ralph Alig

Forest-based activities can involve multiple benefits in addition to contributing to climate change mitigation, such as afforestation on erodible or other environmentally sensitive agricultural land, reducing water pollution, and enhancing wildlife habitat.

and for other ecosystem services under government conservation programs for a project that would be implemented with only one of the funding sources would be socially inefficient. To receive both payments, the landowners would likely need to show that the practices that produce the additional ecosystem services would not have occurred without additional investments. Similarly double counting might occur when a landowner manages for a single ecosystem service but attempts to receive payments for multiple service streams. For example, RFF (2009) suggested that allowing landowners to sell water quality credits from land that is only being managed for carbon would not generate the correct incentives for landowners and undercut the effectiveness of the water quality market.

Baker and Galik (2009) evaluated the impacts of offset payments on the CRP. They found that when lands and projects produce low rates of carbon sequestration, a direct offset market-induced shift out of the CRP is unlikely. In areas that can produce high sequestration rates, however, offset payments could exceed CRP payments at a carbon price of $5.56/t CO_2e. Above this price, implementation of cap-and-trade legislation may induce landowners to drop out of the CRP program. Baker and Galik (2009) suggested three approaches for avoiding this scenario: (1) increase CRP rental payments, (2) alter the Environmental Benefits Index (EBI) to put more weight on carbon storage, or (3) allow CRP contract holders to participate in the carbon offset market. Increasing rental payments or altering the EBI would maintain the basic structure of CRP while adding flexibility to its implementation. However, it would also likely result in large increases in the CRP budget, which may not be politically or fiscally feasible. Allowing CRP contract holders to participate in a federal carbon offset program would resolve some of the budget concerns. Although CRP landowners are currently allowed to sell carbon credits in voluntary markets, a number of complicated issues concerning additionality, baselines, and double counting would have to be resolved before CRP landowners would be allowed to sell carbon credits in a federal cap-and-trade system.

Summary and Conclusions

A large literature on carbon offset economics and policy has developed over the past two decades. However, few carbon offset studies concentrate on NIPF landowners. Therefore, we have reviewed the general forest and nonforest carbon offset literature to evaluate the competitiveness of U.S. forest offset projects. From this literature review we make inferences concerning the direction of influence on NIPFs. The cost comparisons reported here reflect studies from the available literature that are not completely comparable to one another in terms of assumptions and methodologies. Therefore, the study findings and cost comparisons should be interpreted with caution as general trends rather than definitive conclusions.

A number of factors influence the potential for forest offset projects to reduce GHG emissions including tree species and site characteristics, management practices, longevity of wood products and disposal methods, opportunity costs of land, discount rates, and forest and agricultural prices. Differences in opportunity cost of land, discount rates, impacts of agriculture sector prices, carbon loss occurring during conversion to wood products and disposal, and biological factors result in wide variation of unit cost estimates. Cost estimates for producing U.S. forest offsets range from about \$2 to \$77/t CO_2e (not including transactions costs or costs to meet regulatory rules concerning additionality, baselines, permanence, leakage, etc.). United States forest offset projects tend to be less costly than European projects but more expensive to implement than those in tropical forests in developing countries.

Afforestation projects tend to be more cost effective than forest management (e.g., extending rotation age) projects. Estimated costs for U.S. afforestation projects ranged from \$1.29 to \$70.92/t CO_2e with an average of \$16.78/t CO_2e and median of \$9.85/t CO_2e. Although the range of cost estimates for U.S. forest management projects was smaller (\$12.16 to \$49.35/t CO_2e), the average and median costs were higher at \$27.65/t CO_2e and 24.54/t CO_2e, respectively. Afforesting pasturelands appears to be more cost effective than afforesting croplands owing to the higher opportunity cost of cropland. The CBO (2007) study concluded that a price of $5/t CO_2e would prompt enough tree planting to sequester between 2 and 50 million metric tons annually and that a price of $50/t CO_2e would increase the amount sequestered to 500 to 800 million metric tons annually. However, others suggest that the lack of availability of lands for afforestation may limit its usefulness in the United States (Gorte 2009a, Plantinga et al. 1999, Ray et al. 2009).

Our review of the literature suggests that the opportunity cost of land, transaction costs, and costs associated with meeting eligibility criteria such as establishing baselines and insuring additionality and permanence are the most important factors driving the costs of forest offset projects. The literature suggests that transaction costs will vary depending on project size and activities (e.g., reforestation vs. extending rotation ages) and could be a significant cost and deterrent for NIPF participation in the offset market. Transaction costs are one of the key challenges that private forest landowners may face in participating in emerging carbon markets. Nevertheless, although transaction costs may be a crucial factor in total project costs, they may dim in comparison to the costs associated with complying with various project criteria and accounting requirements. The most important of these are establishing baselines, additionality, and permanence.

Eligibility criteria in proposed and existing offset programs in the United States differ widely in how they address individual carbon pools, baselines, additionality, permanence, and leakage. As a result, the amount of creditable carbon generated for a project and the required break-even price of carbon offset credits also differ dramatically under the different protocols. Failing to account for the added costs associated with these protocols would likely result in underestimating forest offset project costs.

Although forest carbon offset projects on NIPF lands may be at a competitive disadvantage compared to a number of carbon offset alternatives, there are a range of policy options that might narrow the gap. The most important are rules for determining baselines and additionality. Many of the cap-and-trade proposals require establishing a business-as-usual baseline. However this produces a number of challenging hurdles to forest projects and particularly for NIPF landowners as they must develop a projection of future carbon stocks (in the absence of the project) in the face of major uncertainties concerning future market prices, laws, regulations and policies, and ecological and social conditions. The simpler base-year performance standard approach would likely facilitate NIPF landowner participation, reduce costs, and improve competitiveness (Fenderson et al. 2009). However, it is viewed by many as an inferior approach that might result in significant amounts of "non-additional" tons being credited.

Three options for ensuring permanence are typically offered in most cap-and-trade proposals: insurance, offset reserves, and term offset credits. Upfront costs should be lower under term contracts than requiring landowners to purchase insurance or establish an offset reserve. The Southern Group of State Foresters (Fenderson et al. 2009) suggest that long-term contracts or conservation easements would deter many NIPF owners and that term credits of 10 to 20 years would be most beneficial for NIPF landowners in the South given rotation lengths that range from 25 to 80 years. However, under the Kyoto agreement, TCERs have suffered a price disadvantage compared to "normal" credits.

Finally, stacking of ecosystem services payments or credits with carbon offset payments may be one of the more important issues in improving the participation of NIPF landowners. Although care needs to be taken to insure additionality and avoid double counting and leakage, permitting landowners to profit from producing a range of services including carbon may be necessary for NIPF landowners to be able to compete in future offset markets. However, given the myriad financial incentive programs with multiple and often conflicting goals currently influencing land use, this needs to be done based on sound policy research, as changes to these policies will determine future landowner participation in both carbon markets and traditional land conservation programs. Research is needed to understand how to resolve potential conflicts between government conservation programs and the possible federal GHG offset market to produce maximum participation by NIPFs. Research in this area should endeavor to understand how landowners respond to a wide combination of market and policy incentives and education and community-based initiatives. Knowledge of how multiple incentives with different goals interact and what outcomes, in terms of landowner behaviors, ensue is essential for developing cost-effective and efficient policies.

English Equivalents

When you know:	Multiply by:	To get:
Hectares (ha)	2.47	Acres
Metric tons (t)	1.102	Tons
Metric tons per hectare (t/ha)	0.45	Tons per acre
Dollars per metric ton ($/t)	0.907	Dollars per ton
Dollars per hectare($/ha)	0.405	Dollars per acre

Literature Cited

Adams, D.M.; Alig, R.J.; McCarl, B.A. [et al.]. 1999. Minimum cost strategies for sequestering carbon in forests. Land Economics. 75(3): 360–374.

Adams, R.; Adams, D.; Callaway, J. [et al.]. 1993. Sequestering carbon on agricultural land: social cost and impacts on timber markets. Contemporary Policy Issues. 11(1): 76–87.

Adams, D.M.; Haynes, R.W. 1980. The 1980 softwood timber assessment market model: structure, projections, and policy prediction. Forest Science Monograph 22. 64 p.

Alig, R.; Adams, D.; McCarl, B. [et al.]. 1997. Assessing effects of mitigation strategies for global climate change with an intertemporal model of the U.S. forest and agriculture sectors. Environmental and Resource Economics. 9: 259–274.

Alig, R.J.; Latta, G.S.; Adams, D.M.; McCarl, B.A. 2010. Mitigating greenhouse gases: the importance of land base interactions between forests, agriculture, and residential development in the face of changes in bioenergy and carbon prices. Forest Policy and Economics. 12(1): 67–75.

Antle, J.M.; Capalbo, S.M.; Mooney, S. [et al.]. 2001. Economic analysis of agricultural soil carbon sequestration: an integrated assessment approach. Journal of Agricultural and Resource Economics. 26(2): 344–367.

Baker, J.S.; Galik, C. 2009. Policy options for the conservation reserve program in a low carbon economy. Working paper. Climate Change Policy Partnership. http://www.nicholas.duke.edu/ccpp/ccpp_pdfs/low.carbon.policy.pdf. (June 10, 2010).

Baral, A.; Guha, G.S. 2004. Trees for carbon sequestration or fossil fuel substitution: the issue of cost vs. carbon benefit. Biomass and Bioenergy. 27: 41–55.

Baughman, M.J. 2002. Characteristics of Minnesota forest landowners and the Forest Stewardship Program. Reaching out to forest landowners symposium. Cloquet, MN: University of Minnesota, Cloquet Forestry Center. 14 p.

Beach, R.; Pattanayak, S.; Yang, J.; Murray, B.; Abt, R. 2005. Econometric studies of non-industrial private forest management a review and synthesis. Forest Economics and Policy. 7: 261–281.

Bigsby, H. 2009. Carbon banking: creating flexibility for forest owners. Forest Ecology and Management. 257: 378–383.

Birdsey, R.A. 1992. Carbon storage and accumulation in the United States forest ecosystems. Gen. Tech. Rep. WO-59. Washington, DC: U.S. Department of Agriculture, Forest Service. 51 p.

Bliss, J.; Martin, A. 1990. How tree farmers view management incentives. Journal of Forestry. 88: 23–29.

Boyd, R. 1984. Government support of nonindustrial production: the case of private forests. Southern Economic Journal. 51: 87–109.

Brockett, C.; Gerhard, L. 1999. NIPF tax incentives: Do they make a difference? Journal of Forestry. 97(4): 16–21.

Brooks, D.J. 1985. Public policy and long-term timber supply in the South. Forest Science. 31: 342–357.

Butler, B. 2008. Family forest owners of the United States, 2006. Gen. Tech. Rep. NRS-27. Newtown Square, PA: U.S. Department of Agriculture, Forest Service, Northern Research Station. 72 p.

Callaway, J.M.; McCarl, B.A. 1996. The economic consequences of substituting carbon payments for crop subsidies in U.S. agriculture. Environmental and Resource Economics. 7(1): 15–43.

Carbon Catalog. 2010. Find a carbon offset project. http://www.carboncatalog.org/projects/. (July 1, 2010).

Chomitz, K.M.; Lecocq, F. 2004. Temporary sequestration credits: an instrument for carbon bears. Climate Policy. 4(1): 65–74.

Climate Action Reserve [CAR]. 2009. Forest project protocol version 3. http://www.climateactionreserve.org/how/protocols/adopted/forest/development/. (July 1, 2010).

Cohen, M.A. 1983. Public cost-share programs and private investment in the South. In: Royer, J.; Risbrudt, P.; Christopher, D., eds. Symposium proceedings, non industrial private forests: a review of economic and policy studies. Durham, NC: Duke University: 181–188.

Congressional Budget Office [CBO]. 2007. The potential for carbon sequestration in the United States. A Congressional Budget Office paper prepared for the Congress. http://www.cbo.gov/ftpdocs/86xx/doc8624/09-12-CarbonSequestration.pdf. (June 17, 2010)

Consortium for Research on Renewable Industrial Materials [CORRIM]. 2009. Maximizing forest contributions to carbon mitigation: the science of life cycle analysis—a summary of CORRIM's research finding. Fact sheet 5. http://www.corrim.org/factsheets/fs_05/fs_05.pdf. (October 6, 2010).

Creyts, J.; Derkach, A.; Nyquist, S. [et al.]. 2007. Reducing U.S. greenhouse gas emissions: how much at what cost. Executive Report. U.S. Greenhouse Gas Mapping Initiative. McKinsey and Company. 107 p. http://www.mckinsey.com/clientservice/ sustainability/pdf/US_ghg_final_report.pdf. (June 15, 2010).

de Steiguer, J. 1984. Impact of cost-share programs on private reforestation behavior. Forest Science. 30: 697–704.

Deveny, A.; Nackoney, J.; Purvis, N. 2009. Forest carbon index: the geography of forests in climate solutions. Washington, DC: Resources for the Future. http://www.forestcarbonindex.org/. (June 2010).

Dixon, R.K.; Winjum, J.K.; Andrasko, K.J. [et al.]. 1994. Integrated land-use systems: assessment of promising agroforest and alternative land-use practices to enhance carbon conservation and sequestration. Climatic Change. 27(1): 71–92.

Dutschke, M. 2002. Fractions of permanence—squaring the cycle of sink carbon accounting. Mitigation and Adaptation Strategies for Global Change. 7: 381–402.

Fenderson, J.; Kline, B.; Love, J.; Simpson, H. 2009. Guiding principles for a practical and sustainable approach to forest carbon sequestration projects in the Southern United States. Southern Group of State Foresters. http://www.southernforests.org/publications/Guiding %20Principles%20 for%20Forest%20Carbon%20Seq%2008-2009.pdf/view. (June 30, 2010).

Galik, C.S.; Baker, J.S.; Grinnel, J.L. 2009a. Transaction costs and forest management carbon offset potential. http://nicholas.duke.edu/ccpp/ccpp_pdfs/transaction.07.09.pdf. (June 30, 2010).

Galik, C.S.; Mobley, M.; Richter, D. 2009b. A virtual "field test" of forest management carbon offset protocols: the influence of accounting. Mitigation and Adaptation Strategies for Global Change. 14: 677–690.

Gorte, R.W. 2009a. Carbon sequestration in forests. Report prepared by Congressional Research Service (CRS) for Congress. Washington, DC: CRS RL31432. http://www.fas.org/sgp/crs/misc/RL31432.pdf. (June 16, 2010).

Gorte, R.W. 2009b. U.S. tree planting for carbon sequestration. Report prepared by CRS for Congress. Washington, DC: CRS R40562. http://www.fas.org/sgp/crs/misc/R40562.pdf. (June 15, 2010).

Gorte, R.W.; Ramseur, J.L. 2008. Forest carbon markets: potential and drawbacks. Report prepared by CRS for Congress. Washington, DC: CRS RL34560. http://www.nationalaglawcenter.org/assets/crs/RL34560.pdf. (June 17, 2010).

Greene, J.; Kilgore, M.; Jacobson, M.; Daniels, S.; Straka, T. 2005. Existing and potential incentives for practicing sustainable forestry on non-industrial private lands. Final Report, National Commission on Science for Sustainable Forestry. http://www.srs.fs.usda.gov/econ/data/forestincentives/ncssf-c2-final-report.pdf. (September 15, 2009).

Greene, J.L.; Blatner, K.A. 1986. Identifying woodland owner characteristics associated with timber management. Forest Science. 32(1): 135–146.

Haller,T.; Thoumi, G. 2009. Financial accounting for forest carbon offsets and assets. Special report to Mongabay.com. http://news.mongabay.com/2009/1116-haller-thoumi_forest_carbon.html. (June 30, 2010).

Herzog, H.; Caldeira, K.; Reilly, J. 2003. An issue of permanence: assessing the effectiveness of temporary carbon storage. Climatic Change. 59(3): 293–310.

Huang, C.-H.; Kronrad, G.D. 2001. The cost of sequestering carbon on private forest lands. Forest Policy and Economics. 2: 133–142.

Hyberg, B.T.; Holthausen, D.M. 1989. The behavior of nonindustrial private forest landowners. Canadian Journal of Forest Research. 19: 1014–1023.

Im, E.; Adams, D.M.; Latta, G.S. 2007. Potential impacts of carbon taxes on carbon flux in western Oregon private forests. Forest Policy and Economics. 9(8): 1006–1017.

Ingerson, A. 2009. Wood products and carbon storage: Can increased production help solve the climate crisis? Washington, DC: The Wilderness Society. 39 p.

James, L.; Hoffman, W.; Payne, M. 1951. Private forest landownership and management in central Mississippi. Tech. Bulletin 33. State College, MS: Mississippi State College, Agriculture Experimental Station. 38 p.

Jenkins W.; Olander L.; Murray B. 2009. Addressing leakage in a greenhouse gas mitigation offsets program for forestry and agriculture. NI PB 09-03. Durham, NC: Nicholas Institute for Environmental Policy Solutions, Duke University. 12 p.

Johnson, D.W. 1992. Effects of forest management on soil carbon storage. Water Air Soil Pollution. 64: 83–120.

Kilgore, M.A.; Blinn, C.L. 2004. Policy tools to encourage the application of sustainable timber harvesting practices in the United States and Canada. Forest Policy and Economics. 6: 111–127.

Kilgore, M.A.; Greene, J.L.; Jacobson, M.G. [et al.]. 2007. The influence of financial incentive programs in promoting sustainable forestry on the Nation's family forests. Journal of Forestry. 105(4): 184–191.

Kluender, R.; Walkingstick, T.; Pickett, J. 1999. The use of forestry incentives by nonindustrial forest landowner groups: Is it time for a reassessment of where we spend our tax dollars? Natural Resources Journal. 39: 799–818.

Langpap, C.; Kim, T. 2010. Literature review: an economic analysis of incentives for carbon sequestration on nonindustrial private forests (NIPFs). In: Alig, R.J., tech. coord. Economic modeling of effects of climate change on the forest sector and mitigation options: a compendium of briefing papers. Gen. Tech. Rep. PNW-GTR-833. Portland, OR: U.S. Department of Agriculture, Forest Service, Pacific Northwest Research Station: 109–142. Chapter 5.

Lee, K.; Kaiser, F.; Alig, R. 1992. Substitution for public and private funding in planting of southern pine. Southern Journal of Applied Forestry. 16: 204–208.

Lewandrowski, J.; Peters, M.; Jones, C.; House, R. 2004. Economics of sequestering carbon in the U.S. agricultural sector. Tech. Bulletin TB-1909. U.S. Department of Agriculture, Economic Research Service. 69 p.

Manley, J.; van Kooten, G.C.; Moeltner, K.; Johnson, D.W. 2005. Creating carbon offsets in agriculture through zero tillage: a meta-analysis of costs and carbon benefits. Climatic Change. 68(January): 41–65.

Marland, G.; Fruit, K.; Sedjo, R. 2001. Accounting for sequestered carbon: the question of permanence. Environmental Science and Policy. 4(6): 259–268.

McCarl, B.A. 2007. Agriculture in the climate change and energy price squeeze. Part 2: Mitigation opportunities. On file with: McCarl, B.A. Professor of Agricultural Economics, Texas A&M University, College Station, TX 77843-2124.

McCarl, B.A.; Callaway, J.M. 1995. Carbon sequestration through tree planting on agricultural lands. In: Lal, R.; Kimble, J.; Levine, E.; Stewart, B.A., eds. Soil management and greenhouse effect. Boca Raton, FL: Lewis Publishers: 329–338.

McCarl, B.A.; Schneider, U. 2001. Greenhouse gas mitigation in U.S. agriculture and forestry. Science. 294: 2481–2482.

Mercer, D.E.; Cooley, D.M.; Hamilton, K. [In press]. Payments for forest ecosystem services in the U.S. Washington, DC: Ecosystem Marketplace.

Mercer, D.E.; Pattanayak, S. K. 2003. Agroforestry adoption by smallholders. In: Sills, E.O.; Abt, K.L., eds. Forests in a market economy. Dordrecht, The Netherlands: Kluwer Academic Publishers: 283–300.

Mercer, D.E.; Prestemon, J.P.; Butry, D.T.; Pye, J.M. 2007. Evaluating alternative prescribed burning policies to reduce net economic damages from wildfire. American Journal of Agricultural Economics. 89(1): 63–77.

Metz, B.; Davidson, O.; deConinck, H. [et al.]. 2005. Summary for policymakers and technical summary. A special report of working group III of the Intergovernmental Panel on Climate Change. 62 p. http://www.wmo.int/pages/ prog/ drr/publications/drrPublications/IPCC/IPCC_CO2_e.pdf. (June 15, 2010).

Mooney, S.; Brown, S.; Shoch, D. 2004. Measurement and monitoring costs: influence of parcel contiguity, carbon variability, project size and timing of measurement events. Arlington, VA: Winrock International. 20 p.

Moulton, R.; Richards, K. 1990. Costs of sequestering carbon through tree planting and forest management in the United States. Gen. Tech. Rep. WO-58. Washington, DC: U.S. Department of Agriculture, Forest Service. 47 p.

Newell, R.G.; Stavins, R.N. 2000. Climate change and forest sinks: factors affecting the costs of carbon sequestration. Journal of Environmental Economics and Management. 40(3): 211–235.

New York State. 1991. Analysis of carbon reduction in New York State. Report of the New York State Energy Office, in consultation with NYS Department of Environmental Conservation and NYS Department of Public Service. New York: NYS Energy Office. 105 p.

North Carolina Solar Center [NSCS]; North Carolina State University; and Interstate Renewable Energy Council. 2010. Database of State Incentives for Renewable Energy and Efficiency [DSIRE]. Database. http://www.dsireusa.org/. (July 1, 2010).

Olander, L.; Profeta, T.; Galik, C. 2010. Sticking points in offsets policy. Nicholas Institute Discussion Memo. Durham, NC: Nicholas Institute for Environmental Policy Solutions, Duke University. http://nicholas.duke.edu/ institute/offsets.memo.01.07.10.pdf. (July 2, 2010).

Parks, P.J.; Hardie, I.W. 1995. Least-cost forest carbon reserves: cost-effective subsidies to convert marginal agricultural land to forests. Land Economics. 71(1): 122–136.

Parry, M.L.; Canziani, O.F.; Palutikof, J.P. [et al.], eds. 2007. Climate change: impacts, adaptation, and vulnerability. Contribution of working group II to the fourth assessment report of the Intergovernmental Panel on Climate Change. Cambridge, United Kingdom: Cambridge University Press. 1000 p.

Pautsch, G.R.; Kurkalova, L.A.; Babcock, B.A.; Kling, C.L. 2001. The efficiency of sequestering carbon in agricultural soils. Contemporary Economic Policy. 19(2): 123–134.

Pendell, D.L.; Williams, J.R.; Rice, C.W. [et al.]. 2006. Economic feasibility of no-tillage and manure for soil carbon sequestration in corn production in northeastern Kansas. Journal of Environmental Quality. 35(4): 1364–1373.

Perez-Garcia, J.; Lippke, B.; Briggs, D. [et al.]. 2005. The environmental performance of renewable building materials in the context of residential construction. Wood Fiber Science. 37(5): 3–17.

Plantinga, A.; Mauldin, T. 2001. A method for estimating the cost of CO_2 mitigation through afforestation. Climatic Change. 49: 21–40.

Plantinga, A.J.; Mauldin, T.; Miller, D. 1999. An econometric analysis of the costs of sequestering carbon in forests. American Journal of Agricultural Economics. 81(4): 812–824.

Ray, D.G.; Seymour, R.S.; Scott, N.A.; Keeton, W.S. 2009. Mitigating climate change with managed forests: balancing expectations, opportunity and risk. Journal of Forestry. 107(1): 50–51.

Resources for the Future [RFF]. 2009. Ecosystem service stacking: Can money grow on trees? Washington, DC: Resources for the Future. http://www.rff.org/wv/archive/tags/Offsets/default.aspx. (June 30, 2010).

Richards, K.R. 1997. Estimating costs of carbon sequestration for a United States greenhouse gas policy. Boston, MA: Charles River Associates. On file with: K.R. Richards, associate professor, School of Public and Environmental Affairs, Indiana University, Bloomington, IN 47405.

Richards, K.R.; Moulton, R.; Birdsey, R. 1993. Costs of creating carbon sinks in the U.S. Energy Conservation and Management. 34(9-11): 905–912.

Rochon, E. 2008. False hope: why carbon capture and storage won't save the climate. The Netherlands: Greenpeace International. 44 p.

Schoeneberger, M.M. 2005. Agroforestry: working trees for sequestering carbon on ag lands. In: Brooks, K.N.; Ffolliot, P.R., eds. Moving agroforestry into the mainstream. Proceedings of the 9th National American Agroforestry Conference. St. Paul, MN: Department of Forest Resources, University of Minnesota: 26–37.

Shaikh, S.L.; Sun, L.; van Kooten, G.C. 2007. Are agricultural values a reliable guide in determining landowners' decisions to create forest carbon sinks? Canadian Journal of Agricultural Economics. 55(1): 97–114.

Siegel, W.C.; Hickman, C.A. 1989. Taxes and the southern forest. In: The South's fourth forest: alternatives for the future. Misc. Pub. 1463. Washington, DC: U.S. Department of Agriculture, Forest Service: 101–121.

Skog, K.E.; Nicholson, G.A. 1998. Carbon cycling through wood products: the role of wood and paper products in carbon sequestration. Forest Product Journal. 48: 75–83.

Smith, D.M.; Larson, B.C.; Kelly, M.J.; Ashton, P.M.S. 1997. The practice of silviculture: applied forest ecology. New York: John Wiley and Sons. 537 p.

Smith, W.B.; Miles, P.D.; Perry, C.H.; Pugh, S.A. 2009. Forest resources of the United States, 2007. Gen. Tech. Rep. WO-78. Washington, DC: U.S. Department of Agriculture, Forest Service. 336 p.

Sohngen, B.L.; Brown, S. 2006. The influence of conversion of forest types on carbon sequestration and other ecosystem services in the South Central United States. Ecological Economics. 57(4): 698–708.

Sohngen, B.L.; Haynes, R.W. 1997. The potential for increasing carbon storage in United States unreserved timberlands by reducing forest fire frequency: an economic and ecological analysis. Climatic Change. 35(2): 179–197.

Sohngen, B.L.; Mendelsohn, R. 2003. An optimal control model of forest carbon sequestration. American Journal of Agricultural Economics. 85(2): 448–457.

Stainback, G.A.; Alavalapati, J.R.R. 2002. Economic analysis of slash pine forest carbon sequestration in the Southern U.S. Journal of Forest Economics. 8(2): 105–117.

Stavins, R.N. 1999. The costs of carbon sequestration: a revealed-preference approach. American Economic Review. 89: 994–1009.

Stavins, R.N.; Richards, K.R. 2005. The cost of U.S. forest-based carbon sequestration. Arlington, VA: PEW Center on Global Climate Change. 52 p.

U.S. Department of Agriculture, Natural Resources Conservation Service. [USDA NRCS]. 2003. National resources inventory: 2003 annual NRI. http://www.nrcs.usda.gov/technical/NRI/2003/national_landuse.html. (June 30, 2010).

U.S. Environmental Protection Agency [USEPA]. 2005. Greenhouse gas mitigation potential in U.S. forestry and agriculture. EPA 430-R-05-006. 154 p.

U.S. Environmental Protection Agency [USEPA]. 2009a. Carbon sequestration in agriculture and forestry. http://www.epa.gov/sequestration/faq.html. (June 30, 2010).

U.S. Environmental Protection Agency [USEPA]. 2009b. Endangerment and cause or contribute findings for greenhouse gases under the Clean Air Act. http://www.epa.gov/climatechange/endangerment.html. (June 30, 2010).

van Kooten, G.C.; Sohngen, B. 2007. Economics of forest ecosystem carbon sinks: a review. International Review of Environmental and Resource Economics. 1: 237–269.

Watson, R.T.; Noble, I.R.; Bolin, B. [et al.], eds. 2000. Land use, land-use change, and forestry. New York: Cambridge University Press. 388 p.

Willey, Z.; Chameides, B., eds. 2007. Harnessing farms and forests in the low-carbon economy—how to create, measure and verify greenhouse gas offsets. Durham, NC: Duke University. http://nicholas.duke.edu/institute/ghgoffsetsguide/ghgexerpts.pdf. (June 15, 2010).

Zhang, D.; Flick, W. 2001. Sticks, carrots, and reforestation investment. Land Economics. 77: 443–456.

Glossary

afforestation—The forestation, either by human or natural forces, of nonforest land.

Conservation Reserve Program (CRP) land—A land cover/use category that includes land under a CRP contract. The CRP is a federal program established under the Food Security Act of 1985 to assist private landowners to convert highly erodible cropland to vegetative cover for 10 years.

cropland—A land cover/use category that includes areas used for the production of adapted crops for harvest. Two subcategories of cropland are recognized: cultivated and noncultivated. Cultivated cropland comprises land in row crops or close-grown crops and also other cultivated cropland, for example, hay land or pastureland that is in a rotation with row or close-grown crops. Noncultivated cropland includes permanent hay land and horticultural cropland.

developed land—In the National Resources Inventory (NRI), developed land consists of urban and built-up areas, as well as land devoted to rural transportation. This is a broader category than the "urban" land use considered in this study. This study has not attempted to model net returns to rural transportation use, so this report focuses only on the urban component of developed land.

forest land—Land at least 10-percent stocked by forest trees of any size, including land that formerly had such tree cover and that will be naturally or artificially regenerated. Forest land includes transition zones, such as areas between heavily forested and nonforested lands that are at least 10-percent stocked with forest trees and forest areas adjacent to urban and built-up areas. The minimum area for classification of forest land is 1 acre (0.405 ha). Roadside, streamside, and shelterbelt strips of timber must have a crown width of at least 120 feet (36.6m) to qualify as forest land. Unimproved roads and trails, streams, and clearings in forest areas are classified as forest if less than 120 feet wide.

Intergovernmental Panel on Climate Change (IPCC)—The IPCC was established to provide decisionmakers and others interested in climate change with an objective source of information about climate change. The IPCC does not conduct any research nor does it monitor climate-related data or parameters. Its role is to assess on a comprehensive, objective, open, and transparent basis the latest scientific, technical, and socioeconomic literature produced worldwide relevant to the understanding of the risk of human-induced climate change, its observed and projected impacts, and options for adaptation and mitigation. For more information, see http://www.ipcc.ch/.

land area—The area of dry land and land temporarily or partly covered by water, such as marshes, swamps, and river flood plains; streams, sloughs, estuaries, and canals less than 200 feet (61 m) wide; and lakes, reservoirs, and ponds less than 4.5 acres (1.8 ha).

land cover/use—A term that includes categories of land cover and categories of land use. Land cover is the vegetation or other kind of material that covers the land surface. Land use is the purpose of human activity on the land; it is usually, but not always, related to land cover. The NRI uses the term land cover/use to identify categories that account for all the surface area of the United States. The six major land use categories considered in this study are (1) cropland, (2) pasture, (3) range, (4) Conservation Reserve Program (CRP), (5) forest, and (6) urban. These uses are described in this glossary.

large urban and built-up areas—These areas include developed tracts of 10 acres (4 ha) and more.

nonindustrial private forest (NIPF)—An ownership class of private lands where the owner does not operate commercial wood-using plants.

National Resources Inventory (NRI)—A statistical survey of land use and natural resource conditions and trends on U.S. non-federal lands. The NRI is led by Natural Resources Conservation Service (NRCS), the Department of Agriculture's lead conservation agency. For more information, see http://www.nrcs.usda.gov/technical/ NRI/.

other rural land—A land cover/use category that includes farmsteads and other farm structures, field windbreaks, barren land, and marshland. Some reports refer to this as NRI minor land cover/uses.

pastureland—A land cover/use category of land managed primarily for the production of introduced forage plants for livestock grazing. Pastureland cover may consist of a single species in a pure stand, a grass mixture, or a grass-legume mixture. Management usually consists of cultural treatments: fertilization, weed control, reseeding or renovation, and control of grazing. For the NRI, it includes land that has a vegetative cover of grasses, legumes, or forbs, regardless of whether or not it is being grazed by livestock.

public—An ownership class composed of land owned by federal, state, county, or municipal governments.

rangeland—A land cover/use category on which the climax or potential plant cover is composed principally of native grasses, grasslike plants, forbs, or shrubs suitable for grazing and browsing, and introduced forage species that are managed like rangeland. This would include areas where introduced hardy and persistent grasses, such as crested wheatgrass, are planted and such practices as deferred grazing, burning, chaining, and rotational grazing are used, with little or no chemicals or fertilizer being applied. Grasslands, savannas, many wetlands, some deserts, and tundra are considered to be rangeland. Certain communities of low forbs and shrubs, such as mesquite, chaparral, mountain shrub, and pinyon-juniper, are also included as rangeland.

residential area—Residential area is the sum of acres in lots used for housing units.Estimates of residential area, urban and rural, are based on data from the American Housing Surveys.

timberland—Forest land that is producing or is capable of producing crops of industrial wood and not withdrawn from timber utilization by statute or administrative regulation. (Note: Areas qualifying as timberland are capable of producing in excess of 20 cubic feet [1.4 cubic meters] per acre per year of industrial wood in natural stands. Currently inaccessible and inoperable areas are included.)

urban area—Nationally, there are two main sources of data on urban area. First, the U.S. Department of Commerce Bureau of the Census compiles urban area every 10 years, coincident with the census of population. Second, the U.S. Department of Agriculture, Natural Resources Conservation Service, publishes area of developed land, including urban components, at 5-year intervals as part of the NRI. Although the U.S. Geological Survey, National Aeronautics and Space Agency, Housing and Urban Development Department, and several local, state, and federal agencies also collect data or conduct special-purpose studies on urban area, the census and the NRI provide the only nationally consistent historical series. Because of differences in data-collection techniques and definitions, the NRI estimates of "large urban and built-up areas" is usually higher than the census "urban area" estimates for nearly all states. The census urban area series runs from 1950, whereas the NRI started providing a consistent series in 1982. Historically, the Economic Research Service (ERS) major land use time series (MLUS) has used census urban area numbers. Prior to 1982, census urban area was the only reliable national source of urban area data available. Since 1945, census urban area has been used in the MLUS time series to maintain a consistent series. For comparison purposes, census urban area is checked against the NRI to help project and interpolate census trends between decennial census years.

urban and built-up areas—These areas consist of residential, industrial, commercial, and institutional land; construction and public administrative sites; railroad yards, cemeteries, airports, golf courses, sanitary landfills, sewage plants, water control structures, small parks, and transportation facilities within urban areas.

Appendix 1: Eligible Offset Projects Listed in 2010 American Power Act (Kerry-Lieberman)

- Methane collection at mines, landfills, and natural gas systems and reduction of methane emissions from non-landfill organic waste streams including manure management, composting, or anaerobic digestion projects.
- Capturing fugitive emissions from the oil and gas sector that reduce greenhouse gas emissions that would otherwise have been flared or vented.
- Capture and geological sequestration of uncapped greenhouse gas emissions.
- Recycling and waste minimization projects.
- Biochar production and use.
- Abating the production of nitrous oxide at stationary sources.
- Destruction of ozone-depleting substances that have been phased out of production.
- Agricultural, grassland, and rangeland practices, including:
  - Altered tillage practices, including the avoided abandonment of conservation practices.
  - Winter cover cropping, continuous cropping, and other means to increase biomass returned to soil in lieu of planting followed by fallowing.
  - Improved management of nitrogen fertilizer use.
  - Reduction in methane emissions from rice cultivation.
  - Reduction in carbon emissions from organically managed soils and farming practices used on certified organic farms.
  - Reduction in greenhouse gas emissions due to changes in animal management practices, including dietary modifications and pasture-based livestock systems.
  - Resource-conserving crop rotations of at least 3 years.
  - Practices that will increase the sequestration of carbon in soils on cropland, hayfields, native and planted grazing land, grassland, or rangeland.
  - Reducing greenhouse gas emissions from manure and effluent, including waste aeration; biogas capture and combustion, improved management or application to agricultural land.

- Forest offset projects, including:
  - Afforestation or reforestation of acreage not forested as of January 1, 2009.
  - Forest management resulting in an increase in forest carbon stores, including harvested wood products.
  - Management of tree crops.
  - Adaptation of plant traits or new technologies that increase sequestration by forests.

- Land management changes, including:
  - Improved management or restoration of cropland, grassland, rangeland, and forest land.
  - Avoided conversion that would otherwise release carbon stocks.
  - Reduced deforestation.
  - Management and restoration of peatland or wetland.
  - Urban tree-planting, landscaping, greenway construction, and maintenance.
  - Restoring or preventing the conversion, loss, or degradation of vegetated marine coastal habitats.

Appendix 2: Costs of Removing Atmospheric CO_2 Through Forest Offset Projects

Forest carbon sink studies	Total carbon	Total area	Cost	Cost	Cost	Location	Project type	Method
	Million metric tons	*Thousand hectares*	*Dollars/ hectare*	*Dollars/ t C*	*Dollars/ t CO₂e*			
Adams et al. 1993	140.00	59.00	442.28	18	4.90	Contiguous U.S.	Afforestation	The marginal cost (averaged across all regions) per ton is calculated using a price-endogenous, spatial equilibrium model of the U.S. agricultural sector known as the Agricultural Sector Model (ASM).
Adams et al. 1999	2023.08	145.60	401.52	29.16	7.95	Contiguous U.S.	Afforestation and forest management changes	The analysis measures cost as the net change in producer and consumer surpluses in markets for forest and agricultural commodities through an optimization model for the U.S. forest and agricultural sectors. Alternative carbon flux targets are examined by constraining the model to find market solutions that allow achievement of the targets.
Baral and Guha 2004	316.75	1.00	18,602.34	63.30	17.25	Southern U.S.	Afforestation and fossil fuel substitution	To evaluate the carbon mitigation potential of afforestation and fossil fuel substitution, bottom-up models based on aboveground tree growth rate, carbon uptake in soil and litter, harvest and storage losses, and energy conversion efficiency are used.
Callaway and McCarl 1996	119.32	29.62	143.39	34.09	9.29	Contiguous U.S.	Afforestation	A modified version of the ASM was used in this analysis.
Dixon et al. 1994	5.98	0.030	180.72	4.73	1.29	Oregon, Utah	Afforestation	Project-based carbon sequestration calculation based on bottom-up study by PacifiCorp.
Huang and Kronrad 2001	0.06	0.00	838.78	44.63	12.16	Texas	Forest management	Optimal carbon sequestration rotation calculation using Faustmann approach for different site index and interest rates.
McCarl and Callaway 1995	243.88	47.39	383.74	72.36	19.72	Contiguous U.S.	Afforestation	Integrates forestry data from the Timber Assessment and Market study (Adams and Haynes 1980) with the ASM.

Forest carbon sink studies	Total carbon	Total area	Cost	Cost	Cost	Location	Project type	Method
	Million metric tons	*Thousand hectares*	*Dollars/ hectare*	*Dollars/ t C*	*Dollars/ t CO$_2$e*			
New York State 1991	0.50	0.80	17.33	29.51	8.04	New York	Afforestation	The carbon reduction cost is calculated based on cost per acre and carbon sequestration potential of forest species.
Newell and Stavins 1999	7.66	2.07	699.79	181.13	49.35	36 counties in Arkansas, Louisiana, and Mississippi	Forest management and afforestation	Econometric estimated parameters of a structural model of land use where changes in alternative land uses are linked with changes in the time paths of CO$_2$ emission and sequestration. The econometric estimates of the costs of carbon sequestration are derived from observations of landowners' actual behavior when confronted with the opportunity costs of alternative land uses.
Parks and Hardie 1995	29.96	6.58	967.26	260.29	70.92	Contiguous U.S.	Afforestation	Engineering cost approach to derive a supply schedule for carbon sequestered in trees planted on marginal agricultural lands in the United States. The schedule is used to develop criteria for enrolling lands in a national carbon sequestration program modeled after the Conservation Reserve Program.
Plantinga and Mauldin 2001	41.55	0.28	5457.40	36.28	9.89	Maine, South Carolina, Wisconsin	Afforestation	An econometric model of land use is used to calculate carbon sequestration potential and cost. Econometric parameters are estimated from data on observed land allocation decisions to quantify the relationship between the share of land in forest and the net returns to forestry, among other land use determinants.
Plantinga et al. 1999	12.80	0.19	4596.33	67.61	18.42	Maine, South Carolina, Wisconsin	Afforestation	Econometric land use of models to estimate the marginal costs of carbon sequestration in the three states.

Forest carbon sink studies	Total carbon	Total area	Cost	Cost	Cost	Location	Project type	Method
	Million metric tons	*Thousand hectares*	*Dollars/ hectare*	*Dollars/ t C*	*Dollars/ t CO₂e*			
Richards et al. 1993	42 903	86.40	3446.72	6.94		Contiguous U.S.	Afforestation and forest management	Engineering or "least to cost" approach is used to develop marginal and total cost curves for the use of tree planting and modified forestry practices to capture atmospheric carbon on marginal agricultural land and forest land in the United States.
Sohngen and Brown 2006	2.28	0.22	1921.95	130.00	35.42	12 states in southern and western regions of the U.S.	Forest management	Engineering cost approach of calculating marginal costs of carbon sequestration in forests. Optimal rotation period with and without terms for the valuation of carbon storage for a range of carbon price for given yield function and forest product price are used to estimate carbon sequestration values.
Sohngen and Haynes 1997	29	198	7.34	50.10	13.65	Continental U.S.	Reducing forest fire frequency (damage)	Linking forest fire mortality model to forest inventory model to determine how changes in the frequency of fires will impact forest inventories. Changes in inventory levels can be used to project both the amount of carbon stored and an economic response.

Source: Adapted from van Kooten and Sohngen (2007). Costs are in 2005 U.S. dollars. The cost estimates in the table are drawn from meta-regression analysis by van Kooten and Sohngen (2007). As studies in their sample provided multiple estimates of one or more projects or regions, the averaged values across a study was calculated by them. $1/t CO_2e = $3.67/metric ton CO_2.

Pacific Northwest Research Station

Web site	http://www.fs.fed.us/pnw
Telephone	(503) 808-2592
Publication requests	(503) 808-2138
FAX	(503) 808-2130
E-mail	pnw_pnwpubs@fs.fed.us
Mailing address	Publications Distribution
	Pacific Northwest Research Station
	P.O. Box 3890
	Portland, OR 97208-3890